The Snow Tourist

The Snow Tourist

CHARLIE ENGLISH

Published by Portobello Books Ltd 2008

Portobello Books Ltd
Twelve Addison Avenue
Holland Park
London
W11 4QR

The journeys in this book have been carbon offset.

A CIP catalogue record is available from the British Library

9 8 7 6 5 4 3 2

ISBN 978 1 84627 083 5

www.portobellobooks.com

Text designed and typeset
in Fournier by Patty Rennie

Illustrations © Barbara Hilliam

Printed in Great Britain
By MPG Books Ltd, Bodmin, Cornwall

For Barbara and Hugh

Contents

CHAPTER I

Snow and Happiness

London

It is late on an autumn afternoon and I am sitting at my desk by the upstairs window which overlooks the blank gable-end of our neighbour's house, making a list of words I associate with snow. There are already several on my notepad. 'Beauty' stands at the top, followed in order by 'danger', 'childhood', 'loneliness' and 'death'. In a separate column, I have written 'sledging', 'skiing', 'snowballs' and 'fun', with two exclamation marks. I am considering putting a line through 'fun' when the streetlight flickers into life. As dusk falls across London, the lamp's orange glow transports me back to February, when snow fell in the steep canyon between the red-brick houses.

On the way to bed that night I stood by the upstairs window watching the snow fall. Flakes burst briefly into the streetlamp's sodium light, to be hustled this way and that by the wind before disappearing into the dark. How many crystals lay down there on the whitening patch of road between our houses? A few million, I guessed. Billions in the rest of the street, and that was a tiny fraction of the number falling from the snowstorm that crossed South-East England. Someone once estimated that a million billion snow crystals were created around the earth every second, in a jumble of shapes and sizes, from simple hexagonal prisms to flat plates and many-fronded stars. How, I wondered, had they calculated that?

In the morning the damp smell of snow permeated the kitchen, now bright with reflected sunshine – what the Japanese call *yuki-akari*, snow light.

My eldest son, Harry, was then almost three, and I wanted to take him outside to show him what had happened. He was smiling as we stepped out on to the white carpet that now covered the garden path. We set off towards the park along the usual route, only this time everything was different. Familiar road markings and patches of gum-spattered concrete were hidden by the newly fallen snow. The parked motorbike under its blue tarpaulin was decorated with ice crystals. Even the sounds were unusual, from the soft crump of our boots to the dampened hum of distant traffic.

We followed the curve of the street, past the café and the postbox, now a brighter red than ever, to the main road, which was empty of its usual queue of vehicles. We walked with short steps along the edge of the park, looking through the railings into the white space beyond. In the winter sunlight, the wet trunks of the horse chestnuts stood out black against the background of snow.

Inside the park, I gathered some snow and squeezed it into a ball – it was sticky and wet. Together, we rolled it along the ground and watched it grow until it creaked under its own weight.

We took a roundabout route home, along Church Street. The banks of TV screens in the window of the electrical shop showed a river of steaming, stationary cars with the caption 'Commuter Chaos'. During the night, the motorways had become long parking lots where people had been forced to sleep in their vehicles. The cold had frozen the points on the railway tracks. Trains had been cancelled. Some of the Underground system was shut, too. But none of that bothered us. We walked on past the bakery and down a back road which the gritting lorries hadn't yet reached. We watched as two vehicles slipped slowly towards each other, wheels locked, coming to a halt with a crunch of impacting plastic. The drivers got out to argue.

Back home we made a foot-high snowman on the table in the back garden. Over the following days it lost its features and shrank until it was just a puddle of cold water and a few twigs.

In the upstairs room, I add two more words to my list: 'innocence' and 'happiness'.

A few weeks before my father died he gave me and my brother copies of a photograph of himself as a young man on skis in an Austrian resort, framed by a bank of spring snow. After his death, we kept these pictures by our beds. I believe this is what he intended: he had already made plans to kill himself, and he wanted me and my brother to remember him by this image of youthful euphoria. The photo was taken before we were born, in the 1960s. He is wearing a chic black ski top with a racing bib tied to his chest. His flip-up shades are down and his teeth bared in a happy laugh, his dark hair parted in a swoop above his face. Even in monochrome, he looks tanned. He was an enthusiastic skier at a time when few people were or could afford to be.

He died when I was ten. Other artefacts have survived the intervening decades, but this image of the skier remains the most powerful. It reveals the joy he found in exertion and speed, in the amplified light and sharp air of the mountains. By all accounts, he was a romantic with the predilection for grand gestures that is the hallmark of those who want things to be simpler than they are. In the end, of course, he chose the blankness of oblivion over life's complicated greys.

I would like to have shared a moment like that with him, but I don't think he ever went skiing again. The camera seems to have caught him at the end rather than the beginning of the race. The fast tilt through the slalom gates is over and he is breathing hard, leaning on his poles, exhilarated. He might not have won, but he has finished, and he is happy.

My first snow arrived a little earlier than Harry's, on the day I was christened. A snowstorm the previous night had blocked the main road into Beverley, the town in northern England where I was brought up.

During the service, a local eccentric came into the church out of the snow. She was known as Umbrella Lil because she always carried a parasol whatever the weather, and she had a reputation for getting into fights and threatening to leave the town for good. She would then hitch a lift to one of the outlying villages, but was always back a few hours later. The congregation inside St Mary's wondered if she'd interrupt and say something like 'This cannot go on', but she remained quiet until shortly after the vicar had completed the baptism, when she announced to my mother that I was a snowflake baby.

It snowed more often in the 1970s. I remember growing up with snowball fights and snowmen, and momentous days when school was cancelled, when we would drag the sledges down from the attic room where they had been hidden for a year. Snow brought out the cameras even then. A Super-8 film shows me and my brother being towed on the back of a sledge to a famous local hill, Granny's Bump. My father is in his wellington boots, red weekend trousers and Norwegian fisherman's jumper. My mother is wrapped in a long padded coat, with a woollen hat. My brother and I wobble about on the sledge and fall off as they haul us along by a rope. Watching it again now, those three flickering minutes give me a sense of warmth and loss, of *nostalgie de la neige*.

The Easter after my father died, my mother took us skiing for the first time. By then we were hard up, so it had to be on the cheap. We drove to Scotland and stayed in an unheated caravan many miles from the mountains, learning to ski in jeans and anoraks. The Cairngorms in winter are magnificent, but prone to extreme, changeable weather. When we returned to the caravan each night my mother would put our gloves and socks under the grill to get them dry.

Physically, those first days on the pistes were punishing. We would get up early, muscles aching from the day before, and clip our feet into unyielding plastic boots. Arriving at the mountain, I would pray that we could find a parking space close to the ski lifts, but we never could. Instead we had to drag our nefarious clobber a torturous quarter of a mile in painfully uncomfortable footwear, with my brother making light of it as he hurried along. There was no stopping to rest and no escape. The distance we had come and the expense incurred made sure of that.

At first this sport seemed beyond me. Every straining muscle was needed to keep the skis from crossing, which would lead to a fall, as would the opposite: the skis doing the splits. Everything seemed counterintuitive – lean forward instead of back; bend your knees instead of standing tall; put your weight on the outside ski in a turn, when it seemed natural to lean in. But it wasn't long before we were flying down the slope, faster than we knew was safe, shouting with excitement, eyes streaming, a great wide, white softness before us, a thirty-mile view hovering at the edge of our vision. At the bottom, panting and strangely hot despite the snow in our hair and up our backs, we would quickly rejoin the queue for the lift uphill.

It was the intensity of skiing that got me hooked. Up there in the snow, all the troubles of the lowlands seemed irrelevant. Sitting in the car on the way home, with the fan heater struggling against the misted-up windows, I realized that for a whole week I had thought of nothing but skiing, the mountain, and snow.

I have returned to the mountains every winter since. In the early days we took the Great North Road – York, Newcastle, Edinburgh – the route of my maternal family's migration south. Scotland seemed to begin only when we had crossed the Forth road bridge, but wasn't really exciting until we'd crossed the Tay and were in the Highlands.

When we could afford it, and as Scottish snow became less reliable, we went to Switzerland, Austria, Italy and France.

One year, as students, a friend and I drove around the French Alps. We spent our evenings drinking supermarket beer and listening to the radio in the front seats of the car, retiring early to sleep in the back so as to be ready to set off to the mountain at first light.

Every autumn now my thoughts return to snow. Snow is something I identify myself with. Like my father, I am a snow person.

My father didn't love snow exclusively. He also loved sailing and the country, and particularly the North Yorkshire landscape he had grown up in. He was raised on the banks of the River Esk, just outside Whitby, the fishing port that was once home to Captain Cook and to the famous father-and-son whaling skippers, both named William Scoresby. In moments of crisis, such as the time he had a heart attack, my father said all he wanted to do was to go back to the Esk to watch the dippers feeding. The river with its birds and wild flowers appeared to him as a utopian homeland where he could be at ease. Like his father and grandfather, he was addicted to that wild corner of England.

The longing for landscape is something I have inherited. When I first arrived in London from what city people call the provinces, I fled back to the countryside at every opportunity. It took me two years to get used to being a Londoner, but the feeling of dislocation never entirely left me. Fifteen years later, after my girlfriend Lucy – soon to be wife – and I had our first two children, London life became more than claustrophobic. It became unbearable. Every day was crowded with people and obligations. We needed to get out, I told Lucy, to somewhere we could find some freedom. I envisaged a simple existence among windswept uplands and immense skies, where I could push the children out of the back door and into a rustic, Arthur Ransome idyll. Lucy saw it differently. While I wasted my life shuttling back and forth from the stockbroker belt, she would be stuck among career housewives organizing coffee mornings. She had come to the capital for a reason. She did not want to go back to provincial life. To her, my dissatisfaction was a state of mind that would move with us. It would be the same wherever we made our home.

The sense of being trapped did not go away. Instead, it gathered momentum so that I began to wonder if these were the feelings my father had felt. Would I deal with them any better? Escape now seemed a priority. It was a question of survival. I could see only one way out: I would plan a trip, an expedition.

I had created 'expeditions' before, to the Sahara, to Pakistan and western China, the Hindu Kush, the Himalayas. At one time in my life, barely a year would go by without an extended bout of travelling. Like many expeditions that pioneered 'the first route' up this or 'the fastest crossing' of that, they were undertaken as much for the benefit of the exploring party as for the objective, and were none the worse for that.

The expedition I decided upon one grey day in London consisted of a series of journeys linked by a single natural form – snow. I would travel to the best snow in the world, discover how people lived with snow, and what they did with it. As on previous expeditions, the principal objective would be the journey itself, the knowledge and experiences I would gather, and the people I would meet along the way.

It was a good time to document snow's story, because it was becoming rarer. Scientists had debated climate change for years, but now politicians were talking about it too. The newspapers were full of stories about rising temperatures. They warned that in some places snow would disappear altogether. Ski resorts were being abandoned or being built ever higher up the slopes. Tourists were travelling to Alaska to see the glaciers while they were still there. Arctic villages were being relocated because the warmer weather was causing pack ice to melt early in the season, eroding the shoreline. The Inuit Circumpolar Conference was suing the US government for infringement of their human rights by not signing up to the Kyoto Treaty on climate change. Every season and almost every month seemed to produce a new record high temperature.

That autumn, I bought an inch-thick wooden board and constructed a desk in the spare corner of an upstairs landing, between the hot-water tank and the bathroom. On a cupboard door I pinned a photomicrograph of a snow crystal and a map of Alaska, and on the opposite side of the desk I cleared some shelves of a heaving weight of novels and once-read travel guides and started to fill them with books on snow. I began with adventures – Apsley Cherry-Garrard's account of travelling to the Antarctic with Scott and Fritjof Nansen's crossing of Greenland by ski – and followed them with books about the Inuit. Weather forecasting guides came next, then books about the influence of El Niño, the science of water, biographies of Kepler and Descartes, as well as titles about avalanches, snow accidents and how to avoid them. Then came philosophy and art – Edmund Burke's reflections on the sublime, Byron's poetry, monographs on Pieter Bruegel and Claude Monet – followed by fiction, nature writing and sport: Arnold Lunn's *History of Ski-ing* and a primer on off-piste routes around the Chamonix valley. There were also boxes of maps, photocopies of newspaper and magazine articles, prints of snow-related downloads from the internet. Snow literature proved so extensive that the books outgrew their shelf space and accumulated in drifts around the house.

Almost by definition it was a selfish expedition. As the itinerary took shape, I knew that the hardest part would be leaving Lucy and the boys. If our relationship was ever strained, it was when we were negotiating over who was able to do what, and when and how we valued each other's time. The conversation went round and round. I would be as quick as I could, I said, which was almost true. I knew parenting was not something you could drop in and out of. I would space the expedition out over two winters. Maybe some of it we could do together as a family. When she eventually said yes, I knew how difficult my absence would make things for her.

After months of negotiation, list-making and packing, the day finally came when the first part of my journey was to begin. That winter morning I kissed my family goodbye, closed the door on the warm house and walked away under a sky the colour of dull lead.

Slowly, I felt the weight of the city begin to lift.

CHAPTER 2

A Night Out with Billy

Iqaluit, Qikiqtarjuaq

Guilt overcame me before I reached the Tube station, but to compensate I had a fistful of plane tickets that in the space of a few days would take me further north than I had ever been, to the Inuit territory of Nunavut in northern Canada. The Inuit have lived above the Arctic Circle, in the highest inhabited latitudes on earth, for at least a thousand years. I was travelling to a place populated by bears and swept by blizzards, where naval expeditions had once descended into cannibalism. I wanted to understand what snow meant to the Inuit, and to learn some of the survival skills they had perfected over a millennium and were now in danger of losing. I also wanted to breathe in as much of the icy romance of this place as I possibly could.

On a cold day in early winter, I landed at Iqaluit on Baffin Island, on the shore of Frobisher Bay. A young Canadian woman named Sophia drove me the few miles across town to the boarding house where I had booked a bed. Beyond the Nunavut parliament building and the shops, houses painted in bright Scandinavian colours climbed away towards a low ridge. That afternoon the town's buildings seemed to be blinking in the reflected light from the sea ice and the white hills, like a colony of seals. Sophia pulled up outside a rust-coloured wooden barn and told me to wait in the warmth of the pick-up's cab while she fought with the frozen door lock. Behind the house I could make out stacks of ice several metres high which storm surges had driven up the beach. Beyond them, a mile out to sea, men were driving dog teams across the solid surface of the bay.

When she had prised open the lock, Sophia beckoned me over and I made a dash through the cold into the house. It was −32c today, −50c with the wind-chill factored in. She handed me a key and told me that if I wanted to go into town I should call a taxi – it was a flat rate everywhere. She jotted down some numbers. I waved her off from inside the triple glazing and watched her drive away up the snow-covered track.

Alone in the house, I switched on the clock radio. A man from Environment Canada was reading a list of places and numbers in an old-fashioned north American accent, his sentences punctuated by long gaps filled with atmospheric pops and whistles so that it sounded as if he were speaking to us from the 1950s. *Baker Lake... minus fifty-five... Clyde River... minus thirty-two... wind-chill warning... Pond Inlet... minus thirty-five... extreme wind-chill... minus fifty-one... Wind-chill warning for Iqaluit... 5 a.m. Clear... minus thirty-five... Nanisivik, severe weather warning... minus forty-five...* I imagined his audience sitting by their sets across almost a million square miles of Nunavut, from Eureka to Rankin Inlet, Kimmirut to Cambridge Bay, listening for just how damned cold it was going to be in the morning. Canada, I had read, is the coldest country in the world.

Nevertheless, I had an appointment further north and needed to test the collection of old and borrowed clothes I had cobbled together and brought from London. I laid them out on top of a pine chest of drawers in my room and wondered if I had brought quite enough. There were three pairs of skiing socks, three pairs of moisture-wicking thermal bottoms, a pair of jeans, some padded ski trousers, three pairs of thermal tops, a fleece, a windproof gilet, a balaclava, a fur-lined hat with earflaps, three pairs of gloves – two inners, with fingers, one pair of mittens – and a pair of goggles. The zip on my ski jacket had been broken for some time and I hadn't got round to getting it fixed. I made a mental note to buy safety pins, then began to put everything on. The felt-lined snow boots, made of rubber and leather and each weighing three pounds, came last. I waddled twice around the house, experimentally, made my way to the front door, and stepped outside.

I stood for a few moments to let the alien temperature sink in before setting off up the track. The weaknesses in my clothing were soon apparent: my backside was growing numb. The skin on my face felt oddly tight; and my breath produced a plume of freezing moisture that grew a beard of ice on my balaclava. But as I climbed I became warmer, the chill and the light gradually giving me a giddy feeling of delight.

The snow underfoot was dry and grainy, unlike anything I had seen before. Just over two metres of it falls each year in Iqaluit – more than in much of the arid Arctic – and what falls stays around. The wind had blown it off the hummocks and into hollows, leaving tufts of exposed yellow grass. It was like walking in the dunes behind a beach, but where the sand had been bleached white.

Sled dogs barked. Here and there the snow was stained by faeces and urine, marking where a dog team had been kept. Outside their houses, Inuit were laying out dog traces to dry, in the way that other men, further south, would be mowing their lawns or cleaning their cars and motorbikes. I crossed a low ridge and headed down into town.

Snowploughs were busy in the wide streets. As it was Sunday, most of the shops were closed. I walked past a school and the office of the *Nunatsiaq News*, and round by the sports centre. At the Frobisher Inn, the caretaker told me the restaurant and bar were both shut, and anyway, alcohol was tightly regulated in Iqaluit, just as it was right across the territory, so I dropped down to the Northmart store and bought a sandwich made with dense white bread and meat that tasted of nothing. The winter freeze meant that the shops were out of fresh food. As there are no roads to Iqaluit, supplies coming in by sea from the south have to last through the freeze until the next ships arrive in the early summer.

The other customers were mostly poor-looking Inuit who smiled at me when I caught their eyes before returning to their quiet, urgent conversations. One of them approached a white man and his young son and muttered something to him. 'Hello, David,' said the white man. 'No, I don't have any money for you today.' The Inuit man moved away. I shoved the last of the doughy bread into my mouth and went to buy calories for the trip north.

Returning to the red house, I found the heating turned up uncomfortably high but no one else had arrived in the hours I had been away. I peeled off my clothes and lay on the bed in the dark, listening to the man from Environment Canada and thinking about the journey ahead.

Pond Inlet... 5 a.m.... extreme wind-chill... minus thirty-five...

Iqaluit, which means 'place of many fish' in Inuktitut, lies on the northern shore of the bay named for Martin Frobisher, the explorer, pirate, knight of the realm, Yorkshireman and all-round scoundrel who discovered Baffin Island for Queen Elizabeth in 1576 and gave the English possibly their first view of the Inuit. Frobisher had been seeking greater reward for his dangerous crossing of the North Atlantic. Like John Cabot before him and many who came afterwards, he had been searching for a northerly sea route to Asia, the Northwest Passage. Instead he found the entrance to the bay that now bears his name, which in his optimism he believed would lead to China, and claimed it for England. He stepped ashore on 18 August 1576 and soon encountered a band of Inuit. Christopher Hall, skipper of Frobisher's flagship, the *Gabriel*, recalled: 'They bee like to Tartars, with long blacke haire, broad faces, and flatte noses, and tawnie in colour, wearing Seale skinnes, and so doe the women, not differing in the fashion, but the women are marked in the face with blewe streekes downe the cheeks, and round about the eyes.'

Initial contacts with the Inuit went well. After five of his men went missing ashore, however, Frobisher took an Inuit hostage for his crewmen's release. When the men still did not return, and after snow fell on deck one August night, he retreated across the Atlantic with his hostage as evidence of his discovery. The Inuit man died shortly after reaching England, but Frobisher was not yet done with the land Elizabeth dubbed 'Meta Incognita', the Unknown Shore. On his return the following summer he kidnapped more Inuit, this time a man, woman and child. They disembarked at Bristol, where the city's mayor wasted no time in persuading the man to demonstrate his kayak and bird-spear by hunting duck on the River Avon in front of a large crowd. 'He would hit a ducke a good distance of and not misse,' one contemporary account recorded. But the three curiosities from the New World did not survive long. 'They died here within a month.'

Frobisher visited Baffin Island three times in all, but found neither the riches nor the route to the Pacific he had sought. Over the centuries, in the wake of his ships came a host of adventurers who would write their names on the coastline of northern Canada. John Davis, for whom the Davis Strait is named, first explored the region in 1585; Henry Hudson discovered Hudson Bay in 1610; William Baffin explored Baffin Island in the early seventeenth century. After them came the whaling captains and missionaries, and further attempts to find the Northwest Passage, such as Sir John Franklin's in 1845, and expeditions to find what happened to Franklin when he didn't return. (All 129 members of the expedition perished, though many survived for several years after their ships became locked in the ice.) Most of these visitors to the Arctic paid little heed to the inhabitants of the land they found, but the few that did were able to acquire a vast knowledge of the environment. Armed with some of this information, explorers like the twentieth-century anthropologist

Vilhjalmur Stefansson made repeated expeditions across the Arctic that each lasted several years.

The Inuit had a host of different uses for snow, and a hundred different tricks for living with it and travelling through it. They made snow platforms to keep meat out of the reach of dogs; snow walls to act as windbreaks when fishing or around their igloos; snowmen for target practice with spears and bows. They made children's toys with it; they used snow as sponges if the igloo roof began to drip, and as a source of water. They made household furnishings from snow: beds, tables, chairs. Best of all, they made their houses from it.

The raw materials in the Arctic were limited: there was bone, oil, hide, moss, meat, ice, stone and snow. From such limited means the Inuit developed a breathtaking range of equipment for hunting and travel. As well as the skin whaling boat, the *umiak*, which could carry twenty people or a tonne and a half of freight, they made the harpoon and the bow, the caribou parka, the hunting kayak for catching smaller animals, the skin snow boots, *kamiks*, which shrug off snow and dry quickly, and goggles of ivory to prevent snow-blindness.

The *qomatiq*, or sled, was built from driftwood or bone, its crosspieces lashed together with gut loosely enough for it to bend and twist over the bumps. Its whalebone runners were much more efficient in this environment than a wheel would have been. They didn't get bogged down or require moving parts that would freeze. Depending on the condition of the snow, the sled might occasionally be hard to pull but the Inuit had ways to deal with this, as they did with almost every condition they encountered. The half-Greenlandic explorer Knud Rasmussen, who spent many years travelling among the Inuit, remembered one of these from his fifth Thule expedition. 'As long as the snow was moist, and the air not too cold, iron or steel runners make quite good going,' he wrote. 'But as soon as it falls below −20c they begin to stick, and the colder it gets the worse it is. The cold makes the snow dry and powdery, until it is like driving through sand, the runners screeching and whining with the friction, so that even light loads are troublesome to move.' The solution was to coat the runners with a paste made from peat softened in water, over which was laid a thick coat of ice. With this ice shoe, even heavy loads would run smoothly in the lowest temperatures. 'It had, of course, been observed that ice ran easiest over snow, and obviously it would be an advantage to give the runners a coating of ice,' wrote Rasmussen. 'This method at once proved eminently successful, and has remained unsurpassed for rapid running with heavy loads, despite numerous experiments made with other materials by various expeditions.'

The engine that drove the sled was the Inuit dog, a shaggy-haired cousin of the wolf which could keep warm even in the coldest temperatures by burying itself in the snow. Unlike the Native Americans further south who used sled dogs in file to break a trail through the heavy snow, the Inuit harnessed their animals in a fan. This was partly because of the much lower volume of snow-fall in the north: attached this way, each dog could get a better grip on the hard, wind-compacted snow than if they had been treading in one another's foot-prints. They were also less likely to drag each other down if they fell through sea ice. Inuit dogs were so efficient that Rasmussen travelled 20,000 miles around the Arctic with the same team. Even after the arrival of snow-machines in the late 1960s, dogs had major advantages over motorized transport. They didn't break down, and if they were starving they could be fed on a weaker member of the team, as could the driver.

From the first millennium onwards, the Inuit migrated thousands of miles from their homeland in what is now northern Alaska along the coastlines of the polar region. By the thirteenth century they had reached Greenland, where they encountered the Viking colony established by Eric the Red. In one of the few surviving pieces of written evidence from the colony, Ivar Baardssøn, a Greenland official, recorded in the mid-fourteenth century that the western Norse settlement had been taken over by *skraeling*, small people who may well have been Inuit. Whether from the increasingly cold climate that swept the hemisphere at the time, or through battles or competition for food with the Inuit, the Norse lost out. The last written record of their settlement dates from 1408.

The invention for which the Inuit are most famed, and the one I was most keen to find out about, was the igloo. It seems ironic that the knowledge of this structure which gave the Inuit such an advantage in winter was one that they had probably copied from the indigenous culture they had supplanted in the Arctic – a race known to archaeologists as the Dorset people, and to the Inuit as the *Tuniit*.

I met Billy on a stopover between flights from Iqaluit to Qikiqtarjuaq, a small island off the north coast of Baffin Island. The old turbo-prop had been delayed by a blizzard, and the shed that served as a terminal building in Pangnirtung was crowded with Inuit who were trying to get home to their communities further north. I must have stood out from the crowd, dressed in all my layers of ski gear while Billy and the other Inuit hung around in baseball caps, jackets and jeans. He introduced himself, and we sat together on the plane. As we flew over the glaciers and snowfields of the Auyuittuk national park he told me the story of his life.

Now in his early sixties, Billy made his living equipping and guiding expeditions of tourists, fishermen and hunters around his homeland. Sometimes he would take big-game hunters from the south out to shoot polar bear, which was very lucrative. Under Canadian law, the Inuit are adjudged to be part of the ecosystem and allowed a quota of endangered game, but the law doesn't specify that an Inuit has to pull the trigger. American and European hunters pay tens of thousands of dollars for the privilege of killing one.

Until the middle of the twentieth century, Billy's family had been nomadic, living in snow houses throughout the winter and in the summer in tents on the edge of Cumberland Sound. In the 1950s the Canadian government had become concerned about the presence of the itinerant population in the north of their modern state, particularly with how to administer healthcare and education to them, and encouraged the Inuit to settle in camps. Billy's group had settled around the Anglican mission at Pangnirtung. Not long afterwards, the Royal Canadian Mounted Police had slaughtered all the Inuit dogs. The Mounties said they were infected and posed a risk of epidemic, but for many of the Inuit it was too great a coincidence that when the government wanted them to stop moving around they took away their mode of transport.

Billy had been fierce and proud as a young man, with a taste for adventure. He had left Baffin Island to work on the ships supplying the oil rigs off Newfoundland. It was hard work but compared with the Germans, Scandinavians and Americans he was tough and almost impervious to the cold. He was the only Inuit man any of the ships' companies had ever met. One year, while on shore leave, he had visited relatives in Qikiqtarjuaq, the island community near an American base that was part of a network of Arctic stations built during the cold war to warn of Soviet missile launches. There he met a strong-minded local girl and eventually persuaded her to marry him. They have lived together in Qikiqtarjuaq ever since.

That afternoon I went back with Billy to his raised wooden house overlooking the harbour. The streets of Qikiqtarjuaq were made from compacted snow, and far out towards the horizon I could make out an iceberg that was locked in the frozen sea.

We ate with Billy's eldest son, Gary, who was in his early twenties. Billy said he was the most aggressive of his three boys, a natural leader with a tendency to get into trouble – just as Billy had done at that age. As we sat talking in the tremendous heat of the house, Gary didn't seem at all aggressive, his wide smile showing a perfect row of gleaming white teeth. But it was clear that he had led an eventful life. He told me that he had been addicted to ten different types of drugs, including alcohol. Instead of finding happiness, he had found himself needing to earn a large amount of money to fund his habit. Recently, he had given all that up and become a born-again Christian, and his life had been filled with joy. While we talked, he would open his arms and announce,

'Now I have found Jesus! And he has filled me with sweetness and light!' Billy, a lay preacher in the Church, remained impassive.

The conversation turned to igloos. Gary had built his first one when he was twelve. In his grandfather's day it had been a rite of passage. Men who couldn't build igloos well and quickly would be refused permission to marry by the parents of their intended bride. This parental veto applied to all sorts of hunting and snow skills. If the groom seemed a dead loss at survival, he was rejected outright. Their daughter's life, after all, would depend on him catching enough meat and building a warm enough house. Gary's grandfather, who had died two years before, had been a champion igloo-maker. When he and a rival in the community had competed to build igloos side by side, Gary's grandfather had not only marked out a much larger building than his competitor, he had finished it in an hour. His rival was still building his modest igloo when dusk fell.

Would Billy show me how it was done? Sure. We could spend the following night out on the sea ice, if I was strong enough to withstand the cold. I said I was. He looked at the clothes I was wearing and said he wanted to check my gear.

First he felt the weight of the material in my Gore-Tex snow trousers. 'Thin, eh,' he said, before giving me a vast pair of jogging bottoms to put on top. He didn't like my jacket much, either, without its zip, and gave me his old goose-down coat, essentially a tremendously thick sleeping bag with a hole cut in each end. Then he handed me a heavily padded jacket to wear under it, which had been stitched by his wife.

As it turned out, Billy's wife made most of his clothes. She had made him a new pair of green over-trousers and matching jacket from some canvas she had bought at the store, and put a wolf-skin trim around the hood. Wolf-skin was best because snow and ice don't stick to it, so it doesn't get clogged and grow heavy. The best material of all was caribou-skin, Billy said. He had a parka made of that, but it was only for special occasions.

Even with my new gear, it only took a short snow-machine tour of the island to show just how differently we each coped with the cold. Billy sat in front, without goggles or scarf, in the 40-mph stream of −30c air without apparent discomfort. Despite my balaclava, goggles, hat and hood, and the protection of Billy's bulky shape, after a few minutes I was in pain. When we got back I had two white spots of frost-nip on my cheeks, which Billy soon revived by pressing them with his thumb.

Billy said that when he was young, his parents had trained all their children to be tough. They scolded those who complained of cold, and told them not to warm their hands by the fire. 'We had to work out for ourselves the Inuit ways of keeping warm,' Billy said. Some were even better at dealing with the cold than Billy was, in particular two brothers who would ride around on their snow-machines on the coldest winter days without their hoods.

I had just about scraped through the cold test. Back in the house, Gary gave me a quick lesson in self-defence against polar bears, the way his grandfather had taught him. 'If a bear comes at you, you hold your arm up like this,' he said, adopting a crouched position with his arm and wrist curled in front of his face. The bear couldn't bite your head because its jaws wouldn't open wide enough to eat the forearm, end on. If this ploy wasn't successful, Gary said, I should curl up into a tight ball and keep very still. Wouldn't the bear just biff me across the snow? Not if I did this, he said, performing again the arm gesture and the tucked-up ball.

The next morning we prepared to set out for our night on the ice.

'Do you ever fall in the water, Billy?'

'Never.'

During cold snaps at home we used to play ice hockey with walking sticks and a tennis ball on a village pond that was covered by a layer of ice a few centimetres thick. Where the bushes dragged in the water, the pond didn't freeze properly, and when we went near those parts of the ice, we would hear a sound like a taut wire being struck by a hammer and knew to go no further. The Inuit say that sea ice is different from pond ice because of the salt. It doesn't split along shooting cracks, it simply shatters and down you go.

'How thick is the ice in the harbour, Billy?'

He made a face. 'Dunno,' he said. 'Maybe three metres.'

There was nothing to worry about, then; that ice was thick enough to carry a heavy truck. In the Northwest Territories in winter there is an ice highway that runs for hundreds of miles, 85 per cent of it on lakes and rivers, carrying oil tankers to and from the mining operations in the north. In spring, the highway closes until the next freeze-up.

Billy's youngest son, Raymond, helped us load up the long *qomatiq* with all the gear we would need, including extra fuel for the snow-machine, a toolbox, a first-aid kit, a radio, a tent in case the igloo didn't work out, and a rifle to protect us from bears. Raymond packed it neatly and wrapped it all in a tarpaulin, roping it to the slats of the gear sled in the way Inuit had done for a thousand years in preparation for journeys across the sea ice.

We hitched up the snow-machine and Billy and I rode down the white beach and out over the harbour, turning south around the west shore of the island. Snow lay on the ice in dunes, sculpted into soft waves by the wind, stretching in lines to leeward, so that the grain of the frozen sea ran north–south. The ride was fast and bumpy, with the snowmobile and gear sled bucking on the crests and slamming down on the landings. At Old Broughton, we passed some stone ruins where the men of the Hudson's Bay Company had once run a trading post. Further along, we stopped at a cave which had been carved out by the water, a deep fissure in the vertical slabs of rock. The icicles inside looked like stalactites covered with frothy snow, each sounding a different note when you struck it. The sea had frozen on the rocks and pebbled beach. It was as if we were walking in a photograph of foaming waves.

Near the bottom of the island, the sheltered inland water met the wider sea. In summer, when the ice had melted, opposing currents would show themselves in the turbulence of the waves. This energy was now contained within a frozen lid, cracked in places along fault lines that stretched into the distance. We rode across these at high speed, and continued out to sea. Ahead were pressure ridges where storms had forced the ice crust to break and pile up on top of itself. Frozen slabs three feet thick, cobalt blue because of the salt, were piled in jagged barricades. We wove our way around these, Billy standing tall on the snowmobile to look for a way through the maze of snow and ice wreckage, and then, when he had chosen the easiest crossing, riding the complaining machine and sled over the ice, kneeling on the saddle and pushing with his other leg to stop the whole caravan rolling over.

Some kilometres out we came upon a lead in the ice which ran east and followed it, looking for seal holes. Seals keep gaps in the ice open all winter long so that they can come up to breathe and climb out of the water. Bears and Inuit hunters know that if they find a hole that smells strong and fresh, a seal will eventually appear.

Ahead, along the lead, we saw a black thatch of ravens on the ice.

'A hunter must have killed something,' Billy said.

As we approached, the birds lumbered into the air and flew unsteadily to a pressure ridge a short way off, where they sat and examined us. There was a bloody hole in the lead, and a seal's corpse, now little more than a thick skin from which protruded a rack of ribs. Around this was a halo of pink snow, indented with a thousand raven footprints.

The hunter had waited by the hole until he heard the seal coming to the surface, then he shot and hooked it before the current could drag it under the ice. As he had hauled the animal out, its blood had spilt on to the snow. Then he had slit open the seal's belly and chopped out the meat and most of the ribs, leaving the last few standing like broken white sticks. The birds had not yet picked them completely clean, and the skin, with its stringy layer of white blubber, lay almost untouched.

'Seal meat is good for the cold,' Billy said. 'Better than lean caribou.'

We climbed back on to the snow-machine and rode on until, in front of a particularly tall pressure ridge, we stopped again. Billy pointed to a patch of sky in the distance which appeared to be a deeper shade of blue than the rest. 'Under that is open water,' he said. The polar bears would mostly be out there on the floe edge, hunting where the seals were fishing. It was a tight food chain, with each animal trying to win enough energy from the lower orders to stay alive. The seals ate the fish; the bears mainly ate seals; the Inuit at the top of the food chain ate everything: bears, whales, fish, narwhal, caribou. But the bears were not averse to eating humans too. Thinking of Gary's advice, I wondered what the chances were of us surprising one as we thumped down over these ridges. There were large bear prints everywhere we looked.

When the pressure ridges grew as tall as us, Billy suggested we turn inshore to make camp. It was mid-afternoon and we needed time to build the igloo before it got dark. We rode back towards the land, then turned south until at last we rounded the tip of a small island which would shelter our camp from the north wind.

My brother and I once tried to build a snow house in our front garden after a snowstorm. We shaped a thick foundation and piled loose snow on top of it until we had a low wall about two feet high. By then the snow in the garden had run out, but even with extra from next door we couldn't raise it any higher. The wall tapered near the top and refused to take any more snow. The more we bashed it on, the lower the wall sank. Igloo-building was not as simple as we had thought.

Several cultures use snow architecturally, mainly because it is an excellent insulator. The Siberian Koryak people traditionally made a house of willow branches and skins, covering it with banked snow that was too loose to cut into blocks. Europeans in mountainous areas such as the Alps put raised nodules on their roofs to retain the snow in winter. Mountaineers use snow holes – caves dug out of a snowdrift – to insulate them from the cold and the wind, and military survival courses teach how to improvise snow shelters from snow-drifts or by piling snow on top of, for example, a weather balloon and deflating it, or creating a *quinzhee* – a pile of snow that is tunnelled into and hollowed out. But the Inuit snow house, which historically was only built by the tribes of eastern and central Canada and Greenland, is the pinnacle of snow archi-tecture.

As well as being aesthetically pleasing, the shape of the igloo is highly efficient, providing the greatest interior volume for the least amount of wall area and a low profile to the wind. The hemisphere is also the strongest form that snow can be fashioned into, and the tendency of warmed snow to meta-morphose into something akin to concrete means that igloos grow stronger with time. When a layer of ice has built up on the interior walls the shelter becomes as airtight as a bottle, which is why the builder needs to remember to poke a hole in the roof to breathe through. A well-built igloo can support the weight of a polar bear.

For the Inuit, igloos were not only used as survival shelters and winter homes, they also acted as maternity suites, hospitals, mortuaries and cemeteries. A pregnant Inuit woman's family would build her a separate snow house when her time came near, and she would wait inside for the baby's arrival. If someone became diseased, a small snow house was built with an opening in the back. The patient was carried in, the hole was closed up and a door was cut. If they failed to recover, the house was shut up and the diseased person left alone to die. Whether an Inuit died of disease or accident, if the body was recovered it was buried in a tomb made of snow, vaulted so that the corpse did not touch the sides. If it did, according to their beliefs, the soul could escape through the snow.

Snow for the Inuit was believed to be both a conduit and a habitat for the soul. A traditional myth told to the ethnographer Franz Boas relates the story of Omerneeto and her husband Akkolookjo and how children came to be born. In the old days, Boas was told, women who wanted children would wander through the snow until they found the soul of a new baby. Omerneeto, however, was in the habit of borrowing her husband's boots, which were too big for her so that the legs sagged towards the ground. One day, she forgot to lace them up properly, and a soul grabbed on to her laces, crawled up her leg and entered her womb. Ever since then, women have had to give birth.

There were two principal types of igloo built on Baffin Island: the temporary shelter to accommodate a hunter for a single night; and the semi-permanent family winter home. The missionary Julian Bilby, who travelled to Baffin Island a century ago, found that these larger igloos had reached a high level of sophistication, as the houses contained the equivalents of larders and tack rooms. Creating a winter village of igloos was, Bilby wrote, 'quite a work of art'. While staying with the Caribou Inuit of Hudson Bay, Rasmussen recalled seeing great complexes of snow buildings. On one occasion, he was led by a friend's wife into their snow house, which he found was in reality a group of houses cleverly interlinked. 'It was... a real piece of architecture in snow, such as I had never yet seen. Five huts, boldly arched, joined in a long passage with numerous storehouses built out separately, minor passages uniting one chamber with another, so that one could go all over the place without exposure to the weather. The various huts thus united served to house sixteen people in all.'

In 1888, Boas described seeing among the eastern tribes a building called the *qaggi*, or singing house, which was 'a large snow dome about fifteen feet in height and twenty feet in diameter, without any lining. In the centre there is a snow pillar five feet high, on which the lamps stand.'

In Bilby's time, the Baffin Islanders would begin their winter village as soon as the snow lay deep enough, when by common consent all the hunters would remain in camp and join with the old men and boys in building the winter dwellings. The community would work together to build each house, starting with the headman's igloo. Having chosen the site, the builders took sealing spears, long snow knives and saws, and began testing the snow in every direction to find where it was sufficiently deep and wind-packed to be cut into blocks.

Once the hunters had cut an opening in the side of the house, they began making furniture for the igloo. The largest item was a snow sleeping bench, on to which side-pieces were built to form shelves for lamps and utensils, as well as a larder or storage place for oil and blubber. Next they made the porch, which helped keep wind out of the larger room and gave the occupants somewhere to store surplus meat and blubber, dog harnesses, whips and sealing lines. A passage was built leading away from this room to further protect the building from the wind. Then a square was cut out of the dome of the main chamber, and replaced with a thick pane of freshwater ice, which was sealed in place with half-melted snow and then water, to form a window. If lake ice were unavailable they would use strips of dried seal intestine, stitched together.

Finally, the interior would be glazed by blocking the doorway and every entrance, and leaving two lamps burning inside. This raised the temperature so that the surface of the snow melted and refroze to glass-like smoothness. 'The dwelling is proof against draught as if the inside of a bottle,' wrote Bilby. Water would then be thrown on the floor to make it 'smooth as marble and durable as cement'.

When the building work was completed, the women would dress the home, spreading heather on the sleeping bench, followed by heavy winter caribou-skins. Soapstone lamps on stands were placed at each end of the sleeping bench, with a rough framework of wood and deer thongs above them as a rack to dry clothes. Once it was sealed, a traditional snow house would be so warm that even naked Inuit children would get too hot, and have to run outside to cool down. The igloo would be the family's house until the spring thaw, when they would move into a tent.

Building a good igloo requires good snow, and Billy had been anxious about the snow at our camp from the beginning. Even though the winter so far had been very cold, there had been little precipitation, and now in the deepest drift we could find, in the lee of a small iceberg that had been carved by a glacier further up the fjord and was now locked in sea ice, there was barely ten centimetres of powdery, dry hoar. The best snow for building, Billy told me, comes from a single storm, a bank containing only one layer, which is therefore less likely to shear. It will also be neither hard nor soft, but gently compressed by the wind into a cohesive mass that would be strong, sticky and deep enough to produce walls of a good thickness. With that kind of snow an expert could easily cut and mould blocks, and make them into an igloo in less than an hour. Unfortunately, the snow we had to work with had been lying around for weeks, perhaps months. Some of it was hard and had bonded to the ice, while some of it was made up of enormous depth-hoar crystals which shattered as I touched them.

It had been a bad winter for snow, Billy said. Strong winds had redistributed it or packed it together very hard. In the past, people would have had a better idea of where the good snow would be, but now that building igloos was no longer essential to survival, such traditional skills were dying out. Even so, we were both keen to have a go and see how far we got.

We marked out a circle two metres across by staking a string in the middle and walking around the radius, stamping down a broad indentation where the foundation blocks would sit. With a knife – not the traditional double-edged Inuit snow knife but something more like a bread knife – we chopped through the snow in the foundations to make sure it was soft and free of lumps that would make the walls unstable. Then we began to build, cutting large blocks out of the deepest snow we could find, using the saw Billy had brought for the purpose. As we pulled the blocks out, they made a squeaking sound like polystyrene.

The bottom layer should be thicker than the rest, said Billy, to provide a strong foundation. You should start with a low one and make the blocks taller as you work around, so that the wall climbs upwards in a spiral. He fitted them together with the long knife, trimming them to ensure that each one sat snugly against the one before it. The blocks were also cut to lean in towards the centre of the igloo, with the angle of lean becoming more pronounced as we built upwards. As he placed the snow bricks, staggering them as a bricklayer would so that no block sat exactly on top of another, I filled in the spaces between them and put snow around the foundations so they would bond to the snow on the ground. Billy trimmed the top of the wall as he built it up so that it curved smoothly round and upwards like a snail shell. When he had three full layers of blocks he cut an arched doorway so that he could continue to work from the inside.

We worked hard into the evening raising the walls, my arms aching from the weight of the snow. Despite our efforts, the bricks were too thin and the roof we'd created began to fall in. At last we had to admit that our igloo wouldn't support the key snow block at the top. We would have to put up the tent.

I climbed inside the almost-completed snow house and lay down. Even though it was small and unfinished, with a rough floor and gaps in the walls, it was comforting to be inside, protected from the wind and polar bears. There were few better building materials than snow, it seemed to me. There was something perfect about its transience. What structure could exist in better harmony with the earth than an igloo, which creates no waste and leaves no footprint?

The tent that Billy's wife had stitched was of an old-fashioned design, with a ridge-pole. Instead of pegs we tied the guy ropes out to five-gallon jerry cans on one side and the sled on the other. Billy draped the flysheet over the top but didn't bother tying it out; I don't think he really believed in it. With the temperature sinking towards −40C, I thought we might have put it up for whatever protection it afforded. Now that we had stopped work I was beginning to feel the cold. It was time to light the cooker, eat and get warm. But try as he might, Billy couldn't get the rusting old stove to work. There would be no hot food, or heating for the tent, which was barely above the outside temperature.

I retreated to my sleeping bag fully clothed, hoping I could warm the bag by thrashing around inside it, but everything was too cold. I lay there shivering, cursing the stove. My balaclava, heavy with the moisture from my breath, had frozen into a solid helmet. Outside I could still hear Billy crunching about, and I threw black thoughts in his direction. What was he doing? I had known the man for two days, and had trusted him with my life. How stupid could I be? When I heard him start the snow-machine, my fear turned to panic. He was going to abandon me in some forgotten spot in the Davis Strait, forty kilometres from civilization. I imagined the long march back. It might have been possible in daylight, but not in the dark. I would get lost and walk till I dropped, or run into a bear, gesturing futilely with my extended forearm. No, my only hope lay in trying to survive the night here, but might not the igloo be warmer, even without a roof?

It doesn't take much to get into trouble in the cold. When one thing breaks down, it tends to take other things with it. I thought back to what Billy had told me about the time, when he was eighteen, he had taken his snow-machine far out on the land to hunt caribou and it had broken down. He had walked in the direction of home for four days and nights before collapsing on the sea ice. He came to with a pain in his chest from the cold, picked himself up and walked a bit further. At last, someone saw him through binoculars. Billy said he had never been so glad to see a snow-machine in his life.

I was on the point of rushing from the tent and hurling myself at the retreating snowmobile when I heard the engine cut out and Billy came back in with a broad smile. He had sliced the side off the naptha can and managed to light a fire by burning oil in the open tin, a twenty-first-century reprise of the traditional seal-oil lamp. It was too dangerous to bring into the tent and gave out little heat, but its flickering orange tongues felt like a small victory. And Billy wasn't done yet. With fingers that were deft despite the extreme cold, he warmed the stove's rubber plunger on the oil fire and kept trying it until at last he found pressure. When the stove was lit, he warmed the rubber seal for the lamp on it, and within a few minutes we had light and even a little heat. He now moved the stove into the tent and put on it the spicy sausages I had bought in

Iqaluit, and dropped some snow into the kettle. Soon the tent was filled with the smell of grilling pork. I gulped down the hot, garlicky flesh as if it was the best Spanish chorizo, swallowing the lukewarm water like fine wine.

Though it was still not late, we lay in our sleeping bags, the stove roaring and filling the tent with a fug of warmth. I turned on to my back and stared at the canvas ceiling, and Billy began to talk.

He spoke about the Inuits' cousins in Greenland, the high Arctic and the western Arctic. He had travelled to meet them all. The Greenland Kalaallit were so proud and fierce that they had admonished Billy and his sons for speaking in English. He had travelled to Siberia to meet the Yupik, and stayed in a filthy Russian hotel where rats crawled around the floor at night. He spoke about his music, how he had written songs that were sung all over the Arctic. For the Inuit music festival in Quebec, they had built the world's biggest igloo, large enough to seat thousands of people.

As I listened to Billy's stories, with a full belly and warm limbs, my eyelids closed and I drifted off to sleep.

While Basque whalers were fishing the Arctic in the sixteenth century, it wasn't until the eighteenth century that American and European boats arrived in great numbers. Then they came for whalebone to stitch into fashionable corsets and oil to fill the streetlamps of London, and set up permanent stations on these ragged shores. Like Billy, many of the Baffin Inuit have some European ancestry. By the turn of the twentieth century, whale stocks were exhausted and the industry crashed. In its place came the fur trade, and with it the Mounties and the missionaries. The long process of recognizing the Inuit right to self-rule in the territory they had inhabited for seven hundred years began, ironically, after the government had forced them into settlements. By 1976, a new autonomous territory was proposed for northern Canada. In 1999, Nunavut, which means 'our land' in Inuktitut, was proclaimed.

The 2001 census recorded 45,000 Inuit living in Canada, half of them in Nunavut. The population was young and growing rapidly, fertility rates were high and life expectancy increasing. Inuktitut was taught in schools; 70 per cent of Inuit can speak it. A new hospital had just been completed in Iqaluit. Revenue from tourism was growing. The quota system, which allowed the Inuit to hunt their traditional prey, extended to the polar bear and the narwhal. The Inuit names, which no one had forgotten, were now replacing European and American names on the maps for the first time.

Along with the benefits of Westernization came difficulties of the sort that Gary had experienced: the kind that arise when, in the space of a generation, people don't need to do what their parents did to survive. For Inuit men in particular, hunting, once the point and purpose of living, was now becoming redundant.

Climate change brought problems too. The Inuit had made their lives on snow and ice, and if it arrived late or thawed early that meant hardship. Travelling was easier in winter, when the land and the sea were solid and the rocks covered with snow, and the animals could spread out across the ice. Billy said that the winter ice was melting much more quickly than it used to, and didn't extend so far out. The permafrost on which some communities were built had become a boggy swamp. The migration patterns and movements of the animals were changing. One of the most obvious signs of the disruption to the natural environment on Qikiqtarjuaq was the growing number of polar bears that were now seen on or near land, and their increasingly aggressive behaviour. Billy had recently counted twenty-seven on the island at one time, and it is only a few miles across. He also thought there were fewer seals, which meant the bears were marauding ever closer to the town. Gary said he had seen polar bears knocking down cabins in order to try to find sustenance.

I asked Billy what people made of the coming of the white man. There were some in Qikiqtarjuaq who thought he shouldn't get involved with them at all, he said. These people told him he was forgetting his culture, and the Inuit principles. They didn't like the things the white man brought with him, like computers and the internet. 'In some ways they are right,' Billy said, 'I am forgetting the old ways. But people need to move on. You can't just stay stuck.' The white man had brought good things, too. Billy could talk to his daughter, who lived a few hundred miles away in Iqaluit, over the internet, and see pictures of her and the children. The flights had also made a great difference. When he was younger, and lived in the high Arctic, there had been a flight every two weeks, and if you missed it or there was a blizzard you were stuck. People would often have to wait for a month before they could leave. Now you were delayed for a few days at the most.

Billy reckoned his Scottish whaling ancestor meant he could understand the white man better than most. Inuit people are very physical, he said. They live in an aggressive world where they have to be tough and hunt, while 'the white man lives much more in his head'.

Inuit culture had been diluted, diminished by the steamroller of television. But the Inuit were adapting and learning to survive in the new climate, just as they had done before.

The hiss of the guttering stove woke me at 2.30 a.m. and I watched it fizzle out. From then on, the night only grew colder. As I drifted off, I would feel a corner of my body freeze, and have to wake up and shake it back into warmth. Then another foot or arm would go. At 6.30, I could stand it no longer. Day was breaking outside, and I got up and ran around the outside of the tent, stamping my feet and windmilling my arms. When Billy rose an hour later he told me in sympathy that he had not slept well either because it had been too cold, but I had heard him snoring gently through much of the night.

The sun was rising sluggishly over the white mountains of Baffin Island, turning the sea ice which stretched in every direction a watery yellow. We loaded up the sled and Billy hammered the snow-machine across the snow at great speed. When we reached the house I could no longer feel my toes. My cheeks were burning and my fingers were numb, my eyelashes frosted with clumps of frozen tears. My knees didn't want to bend as I climbed off the snow-mobile, and in my agony I turned the Arctic day purple: '*Christ! Jesus Christ!*' Billy, who had once worked on the oil-supply ships but who was a religious man, smiled and said nothing.

Inside, his house was hot and full of people: Billy's niece and her husband, their daughter, plus Billy's wife and daughter, and two more children who were also grandchildren, though I didn't establish exactly whose. His mother didn't speak English, but we traded smiles and nods.

While Billy and I had been struggling with our stove, his mother had been cooking caribou meat and fried potatoes. It was the first proper hot meal I had eaten since arriving in Nunavut, and though the meat looked tough, like jerky, it was tender, lean and delicious.

Afterwards, we watched a debate in the Nanuvut parliament on a huge TV screen, as the children ran in and out, playing games with a puppy.

'How was your night on the land?' Billy's wife asked.

'Fantastic,' I said. 'But really cold.'

She made a dismissive sound with her tongue. 'Nobody stays out on the ice at this time of year,' she said. 'That's why we have this network of cabins for hunters all over the island, with heating and stoves.'

'Really?' I said, astonished.

There was general laughter. Didn't I know? Still smiling, they turned back to watch the debate on their enormous TV.

CHAPTER 3

Reflections on the Six-cornered Snowflake

Jericho, Vermont

Wilson Bentley had fascinated me ever since I discovered his book *Snow Crystals* gathering dust in a London library. Published in 1931, the book contains Bentley's life's work: some 2,500 crystals that fell from the sky and were gathered and photographed down a microscope over a period of fifty years. Bentley begins *Snow Crystals* with hexagonal forms, each equally regular but with different detailing, some like dinner plates, some whose internal symmetries look like something seen in a kaleidoscope. Then Bentley shows us snow flowers, some densely packed like six-petalled roses, others broad-leafed and spread wide like poppies. Then we see shapes like forts, with a hexagon at each corner for added strength. The crystals become more elaborate as the book progresses, as if Bentley is leading us towards the crescendo of the finely fronded stellar dendrites, whose every branch produces six or even twelve more branches, and each of these divides out again.

To Bentley there was no better place for observing snow than the wooded

Green Mountains of Vermont, where he was born and lived. If Nunavut had taught me what snow meant to the culture that had most experience of it, Bentley's country would bring me closer to the science of snow.

There was no point returning to London from Canada. It made better sense to fly straight to New York and drive north to New England. Besides, I was just beginning to enjoy myself. But I could sense the chill in England when, from my hotel in Ottawa, I revealed my plan to go to Vermont.

'Listen,' I said to Lucy, searching for a compromise, 'it's half-term next week. Why don't you all come out?'

As soon as I put the phone down I bought their tickets to New York on the internet. They would be arriving in four days. That, I reckoned, left just enough time for me to reach the Green Mountains.

The history of Western snow science can be said to have begun in the first years of the seventeenth century, when Johannes Kepler, Imperial Mathematician to the Holy Roman Emperor, Rudolf II, crossed the Charles Bridge over the Vltava River in Prague on a snowy afternoon. 'Just then,' Kepler wrote afterwards, 'by a happy chance water-vapour was condensed by the cold into snow, and specks of down fell here and there on my coat, all with six corners and feathered radii.' In similar circumstances, most people would have dusted themselves down and walked on, but Kepler was one of the more enquiring people of the age: among his many achievements he discovered the laws of planetary motion, explained how the moon governs the tides and coined the word satellite. The symmetry of the tiny white stars on his coat engaged his phenomenal intellect, and he soon began a witty treatise on the subject. He gave the resulting pamphlet, *On the Six-cornered Snowflake*, to his patron John Matthew Wacker in 1611 as a New Year's present. Socrates might have dwelt on how far a flea could jump, Kepler wrote, but 'our question is why snowflakes in their first falling, before they are entangled in larger plumes, always fall with six corners and with six rods, tufted like feathers?' The answer, he surmised correctly, lay in the arrangement of the sub-particles which form the snowflake, but he did not have the resources to go much further.

The earliest known observation of the six-pointed structure of snow was recorded by the Chinese scholar Han Ying around 135 BC. Six, according to classical Chinese wisdom, was the symbolic number for the element Water, whereas five was associated with Earth. In the twelfth century the philosopher Zhu Xi wrote: 'Six generated from Earth is the perfected number of water, so as snow is condensed into crystal flowers, these are always six-pointed.'

European philosopher-scientists were a little slower on the uptake. In the thirteenth century, the Dominican friar Albertus Magnus wrote in his commentaries on nature that he had observed that snow crystals were star shaped. Then in 1555, an epic work on northern phenomena produced by the Swedish cleric Olaus Magnus contained a woodcut showing twenty-three fantastical types of snowflake falling out of a beautifully stylized cloud, among them crystals that look like hands, eyes, birds, the crescent moon and even a bell.

Kepler's treatise sparked a European scientific interest in snow. In the winter of 1635, René Descartes made detailed drawings of six-pointed stars which are regarded as the first scientific records of snow crystals. Descartes wrote of his amazement at the minute detailing he found in a shower of stellar dendrites: 'What astonished me most was that among the grains that fell last night I noticed some which had around them six little teeth, like clockmakers' wheels... but so perfectly formed in hexagons, and of which the six sides were so straight, and the six arms so equal, that it is impossible for man to make anything so exact.'

Descartes' work was followed in 1661 by the Dane Erasmus Bartholomew, who wrote an essay suggesting that the 'six-pointed-ness' of crystals resulted from an organized array of smaller elements. In 1665 Robert Hooke sketched a number of snow crystals in *Micrographia*, from observations made with the newly invented microscope.

The first attempts to explain the crystals' different forms began in 1671, when Friedrich Martens, who sailed from Spitzbergen to Greenland as a ship's barber aboard a whaler named *Jonas im Walfisch* (Jonah in the Whale), noted for the first time that snow-crystal types altered with weather conditions. In 1681 Donato Rossetti collected sixty drawings of crystals and tried to group them into distinct types. But the first serious classification system is credited to another whaler, William Scoresby Jr, whose *An Account of the Arctic Regions: With a History and Description of the Northern Whale-fishery*, of 1820, meticulously renders portraits of crystals sketched on the deck while fishing in the Greenland Sea. This won him entry to the scientific establishment through the patronage of Joseph Banks, the explorer and botanist who had sailed the Pacific with Captain Cook and was now president of the Royal Society.

Until the mid-nineteenth century the burgeoning science of snow observation had caused little academic argument. In was then, however, that a row over the properties of snow and ice saw two of the Titans of Victorian science pitched against each other. In one corner was Michael Faraday, who discovered the principles of electricity generation, and his ally, the glaciologist John Tyndall; in the other were the Belfast-born Thomson brothers, James and William. William would go on to become Lord Kelvin, whose name we associate with the scientific scale of 'absolute' temperature. The subject of the controversy was of the utmost urgency and significance: why do snowballs stick together?

In 1842, Faraday embarked upon two decades of investigation into the properties of snow and ice. 'When wet snow is squeezed together, it freezes into a lump (with water between) and does not fall asunder as so much wetted sand or other kind of matter would do,' he wrote in his diary. 'On a warm day, if two pieces of ice be laid one on the other and wrapped up in flannel, they will freeze into one piece.' Faraday and Tyndall suggested that the surface of ice is covered in a thin liquid layer. Provided the temperature of the ice was not too far below freezing, Faraday said, a film on the surface of the snow would remain. But when two pieces of ice were brought together, the point at which the ice blocks joined was no longer a surface, so they froze.

The Thomsons disagreed. They knew that the density of ice is lower than that of water, and predicted that ice would thereby melt under pressure. This explained the welding of two blocks of ice – and the snowball. The ice melted under the pressure of being squeezed, and when the pressure was released it would refreeze.

Posterity sided with the Thomsons. In 1887 John Joly suggested that pressure-melting explained the slipperiness of ice. In the case of the skater on the pond, the weight bearing down through the skates to the ice would melt it, and the water would act as a lubricant. But Joly warned of a major limitation in the theory: the pressure applied by a human skater was not great enough to lower the melting point of the ice by more than a couple of degrees, and skating was clearly possible in much colder temperatures. In 1939 Frank Bowden and T. R. Hughes came up with a new proposal, frictional surface melting, which they then proved by experimentation. The friction between the skate and the ice generated enough heat to provide a microscopically thin cushion of liquid on which the skater could move.

Frictional melting theory still governs winter sports today, as well as providing a convincing answer to such questions as why cars slide on ice. The geophysicist Samuel Colbeck, who has worked extensively on the theory of skiing, calculated that a skier generates enough heat through the ski base to melt a shallow layer of water a millionth of a metre thick over which the skis then glide. According to Colbeck, the material of the ski base, the wax, and even the colour of the ski can affect heat and therefore its speed. Unlike pressure melting, the effect of the 'rubbing' and melting snow can occur even at very low temperatures, but there is a limit. This is why in conditions of extreme cold, such as those experienced by Rasmussen with the Inuit, sleds and skis simply stop working. As Fritjof Nansen related during his trip by ski across Greenland: 'The severe cold we experienced made things... unusually bad; the snow, as we were fond of saying, was as heavy as sand to pull upon.'

Many physicists accept that Faraday's liquid layer exists on ice just below freezing point, although Colbeck is not one of them. The argument over what makes a snowball sticky rumbles on, but Colbeck believes he has the answer. When he stuck a thermometer into a snowball and squeezed it, he observed a drop in the snowball's temperature – clear evidence, he concluded, of pressure melting as proposed by the Thomsons. 'If you put pressure on an ice-grain/water mixture, the pressure is magnified at the contacts between the ice grains,' Colbeck told me. 'To follow the line on the phase diagram, as the pressure increases the temperature of the mixture must decrease. That's what I see when I stick a temperature sensor in a wet snowball and squeeze the snowball. The grains melt at their contacts and that meltwater refreezes when I release the pressure.'

After Faraday and the Thomsons came a new observer of snow, perhaps the most dedicated in history, with a new piece of equipment. He was not a scientist, but a farmer. His name was Wilson Bentley, but he became known as the Snowflake Man.

'You've been upgraded,' said the large woman in charge of the rental lot at Newark airport. It was a sunny, dry, cold morning and I was standing eyeball-to-headlamp with one of the largest cars I had ever seen.

'I booked the saloon. Don't you have anything smaller?'

'This is the last one on the lot.'

It was a black, four-wheel-drive Chevrolet monster weighing two tons, with seven inside seats wrapped in cream calfskin. I placed my few items of luggage in the cavernous boot, engaged the gears and gingerly made my way out of the lot, negotiating the several lanes of the exit barriers towards the open road, like a super-tanker leaving port. I pointed the vast bonnet north, and drove in heavy traffic up I-91 past the built-up cities of western New England, New Haven, Hartford, Springfield, Greenfield. I stopped to eat in a restaurant overlooking the Connecticut River in Brattleboro, where Kipling once lived, then pushed on up to Burlington, a university town with fashionable bars and Bach piped over loudspeakers on the pedestrianized main street. I stayed there a night, and in the morning left for Jericho, a group of settlements just past a Christmas-tree plantation on a road out of Burlington to the north-east. This was Bentley's landscape: a pretty, green place of tumbling streams and low, wooded hills.

Bentley was born in the Mill Brook valley, just east of Jericho, in 1865 and lived there all his life, in a large house with windows through which he could see Mount Mansfield, the highest peak of Vermont's Green Mountains and the only place in the state where snow lingers all year round. In his early teens Bentley spent some of his first earnings on a telescope in order to look at the night sky, but soon he would devote himself to a much smaller type of star. On his fifteenth birthday his mother gave him an old microscope she had used in her days as a schoolteacher, and young Wilson immediately took to it. He put everything he could find in front of the lens: fragments of stone, feathers, petals, drops of water – and snowflakes. Many years later he told a journalist: 'From the beginning, it was the snowflakes that fascinated me most. The farm-folks up in this north country dread the winter; but I was supremely happy, from the day of the first snowfall – which usually came in November – until the last one, which sometimes came as late as May.'

When Bentley was seventeen, his mother persuaded his father, who was deeply sceptical about his son's interest in snow, to spend the vast sum of $100 on a bellows camera that he could attach to his microscope to take photographs of what he saw. After more than a year of experimentation, on 15 January 1885, Bentley became the first person in the world to photograph snow through a microscope. When he first saw that the picture had come out, he felt like falling on his knees beside his camera-microscope and worshipping it.

Today in Jericho Village there is a small Bentley museum, consisting only of a couple of rooms in the Old Red Mill, a large building near a bridge over the narrow river. The mill had been sold to the Jericho Historical Society at the princely price of $1. It was closed when I arrived, but the society's archivist, Ray, agreed to open it and show me round. The walls were lined with Bentley's snow-crystal photographs. He took three-inch plates, Ray told me, and then scratched the black emulsion off the negative in a circle around the image of the crystal so that they appeared as luminous white shapes on a field of black. In a large glass case in one corner of the museum was Bentley's microscope-camera apparatus, with a pair of woollen gloves of the sort he would have worn placed on top. The old $100 bellows camera extended a couple of feet and the brass-cylinder microscope was attached to the front. Connected to the camera was a series of pulleys and strings that allowed Bentley, standing by the glass plate at the other end, to adjust the camera's focal length. He used woodscrews as weights to keep the strings and pulley system in tension. Ray explained that

Bentley was quite a short man, not much more than five feet tall, so he couldn't reach the end of the bellows camera to do the focusing as might a taller person – hence the system of strings and pulleys. The apparatus sat on a wooden table whose legs had been shortened at one end so that the camera pointed slightly upwards, towards the light from the window in Bentley's woodshed.

Seeing the woodscrews hanging from this amateur contraption gave me a pang of pathos. It must have been difficult for Bentley to convince his peers and neighbours of the value of what he was up to in his shed, never mind the father and brother who had disapproved of his experiments from the start. The people of Jericho thought him eccentric, perhaps mad. Even as late as 1925, when he had become a minor celebrity, he told a journalist that his neighbours 'always believed that I was crazy, or a fool, or both. Years ago, I thought they might feel different if they understood what I was doing. I thought they might be glad to understand. So I announced that I would give a talk in the village and show lantern slides of my pictures. They are beautiful, you know, marvellously beautiful on screen. But when the night came for my lecture, just six people were there to hear me. It was free, mind you! And it was a fine, pleasant evening, too. But they weren't interested… I realized then that my hometown folks didn't have a very high opinion of me… They still think I'm a little cracked. I've just had to accept that opinion and try not to care. It doesn't hurt me – very much.'

His neighbours' indifference may have spurred Bentley on; it certainly didn't hold him back. In 1920, articles about him were published in the *New York Tribune*, the *Boston Herald* and *Leslie's Weekly*. On 2 January 1921 the

Boston Globe pronounced: 'Fame Comes to Snowflake Bentley After 3 Patient Years.' *American Embroiderer* followed up by showing its readers how to stitch the snow-crystal forms that Bentley had discovered. He was featured in *Scientific American* and *National Magazine*, and published three articles in *Pearson's Magazine* in London. The New York–based Bray Studios even made a short film with Bentley, entitled *Mysteries of the Snow*. Much of the coverage of Bentley carried in large type a version of his assertion that he had never found two snowflakes alike, and it is largely down to him that the world believes every snowflake is different today. Despite his growing celebrity, his discoveries did not make him rich. When American journalist Mary B. Mullett knocked on his door in 1925, she noted that this was not the home of a wealthy man. The house may have been run down, but finances meant little to Bentley. 'I can't remember the time I didn't love the snow,' he told Mullett. 'I wouldn't change places with Henry Ford or John D. Rockefeller for all their millions. I have my snowflakes.'

A few miles from Jericho Village, Jericho Center has a pretty green and a sign informing passers-by on Route 15 about its most famous former resident. I pulled up outside the store, a white clapboard building which, with its old-fashioned gas pump and stacks of firewood, might have supplied the Waltons. This was where Bentley had bought supplies in the early twentieth century. It didn't look as if it had changed much since. Directly opposite was the town library, with a bell on the roof and two flags outside, one Old Glory and another that read 'Open'. It had been built in the 1820s on the other side of the green, but when they drove the main road through in the 1950s the women of Jericho protested against its demolition, so the townspeople had dragged it across the green to where it now stands.

The door at the top of the wooden, snow-covered steps was on the latch. Inside, there was a little stand displaying the few books about Bentley that are in print. I asked the librarian, Emillie, if they had many visitors come to ask about Bentley. Yes, they got a few visits from strange people who had come here to find out more about him, she said, looking at me closely, gauging whether I was one of them. It turned out that I should have been here last week, when they had held a birthday party for Bentley. Children from the local school had come, and Old Snowflake himself had even made an appearance, or at least one of the teachers dressed up as him.

A friend of Emillie's now lived in the old Bentley residence a couple of miles away on Nashville Road. Emillie showed me a picture of the house in one of the books on the table. The porch on one side had been knocked down, and the front garden was full of trees. In Bentley's time, the trees in Vermont had all been chopped down for firewood, she said, but they had grown back since then.

I drove out of Jericho Center on to the unmetalled Nashville Road, which was corrugated by the tracks of cars that had sped along it. I missed the Bentley place at first, driving several miles up the valley before stopping to ask directions of a man in a rusting Volvo. The house had a snowflake in the gable-end, he said, but as he was driving that way he offered to point it out. In convoy we returned a mile back down the road, speeding along with dust flying, before the Volvo stopped and the driver gestured ahead of us. The trees were thick in the front garden, and the house itself was painted in unexpected mauve and tangerine, with green doors and windows. 'Go on,' said the man in the Volvo, urging me forward. 'Ring the doorbell!' He said he knew the owner. As I was from out of town, she wouldn't mind a bit.

The Volvo sped off and I walked up the snow-covered driveway. There was a double garage, like two barn conversions, with a sign that read 'Oil Deliveries — Please Use Other Door' and a big, rusting cylinder that acted as a doorbell. It boomed out, but nobody came. I looked in through the porch window and saw some cut wood and some craftwork, and got the impression that Emillie's friend was an artist. There were bicycles in the garden, and a shed full of more chopped wood. The house itself looked large enough for several families. Bentley, who never married, had shared it with his brother and his wife.

I walked around outside for several minutes, wondering what it would be like to live here, half hoping the artist owner would return, half hoping that she wouldn't. It was misty and snowing lightly. Up the valley, I could see the looming mass of Bolton Mountain wreathed in grey cloud. I imagined Bentley striding around here, collecting rocks and lumps of quartz to put under his microscope. Ray had told me that Bentley thought Mill Brook Valley the best place in the world for snow. He occasionally travelled to other places, to Buffalo and Montreal, but never found the snow as good as it was in Nashville Road. The valley trapped clouds a little longer here, he said.

After spending so many long days in the unheated woodshed with his snowflakes, Bentley's health eventually gave out, but not before he had won international acclaim. With the publication of *Snow Crystals* in 1931, Bentley, once the outsider, became one of the scientific establishment. But the book would be his memorial. Shortly after receiving the first copies, the 66-year-old made a trip to Burlington and insisted on walking part of the way home. It was cold, with the snow turning to slush in the road. Bentley caught pneumonia and died shortly afterwards, two days before Christmas.

After his death, obituaries appeared in all the national papers. The local *Burlington Free Press*, however, captured him best: 'He saw something in the snowflakes which other men failed to see, not because they could not see, but because they had not the patience and the understanding to look.'

Odd enough not to fit in, determined enough not to care, unconcerned about financial reward, Bentley had pursued the thing he loved until the end. They eventually proclaimed him a genius. I think of him more as a hero.

As I drove back down I-91 from Vermont to meet Lucy and the boys at the airport, I thought about the other snow-obsessives who had followed in Bentley's footsteps and whose discoveries and theories occupied my bookshelf at home. Bentley may have created wonderfully sharp portraits of snow and brought the detail of its crystals to a mass audience, but he was little closer to knowing why they form such shapes than Kepler or Descartes had been two hundred years before. After Bentley, the greatest step towards scientific understanding of the crystals was taken by a Japanese nuclear physicist, Ukichiro Nakaya, who read Bentley's book as soon as it came out and began his own work on snow the following year. Nakaya had found a job at Hokkaido University in northern Japan, where there was no nuclear physics laboratory for him to work with. There was, however, a great quantity of snow, which Nakaya took to studying. After photographing and identifying more than forty different types of snow crystal, he devised a method of growing snow in his lab, and produced the fullest account up till then of the conditions that form many types of crystals.

Nakaya emulated the conditions inside a freezing cloud by streaming moist air through a chamber in a room that could be refrigerated to −30c. Inside the chamber he suspended a fragment of rabbit's hair, which he had found to be the best seed for crystal growth. Nakaya's results showed, if proof were needed, that Bentley's six-sided stars and plates were just a few of the many varieties of snow that exist. He grew prisms, columns, needles, triangular crystals and twelve-branched stars, as well as irregular shapes and *graupel*. He produced a classification system that recorded forty-three different types of snow crystal, and the temperatures and pressures at which the various shapes formed.

Nakaya placed the crystal types on a chart that plotted temperature against moisture content in the air. For example, at −2c and with low moisture content, crystals form into plates, with little variation. At −5c, crystals form columns. At −15c, plates return, but they are particularly large and thin. As the super-saturation of water in the air increases, the crystals become pronged and fluffier, more fernlike.

Using Nakaya's work, physicists are now able to answer most of the questions posed by Kepler, such as, why do snowflakes always fall with six corners and six rods, tufted like feathers?

As Kepler deduced but couldn't know, the near-perfect hexagonal essence of the snow crystal lies in the shape of the water molecule and the way that several molecules bond together. Each molecule of H_2O can form hydrogen bonds with four of its neighbours, which produces a crystal lattice that, when viewed in one plane only, is six-sided.

The simplest form of snow crystal is the hexagonal prism – the shape you would get if you chopped a piece out of the middle of a pencil. Depending on the moisture and temperature conditions, the crystal will tend to grow either along the length of the pencil, stretching the six prismatic planes, or outwards from the lead of the pencil (along the basal plane). The first of these types of growth will produce needles and columns, the second flat hexagonal plates. When they reach a certain size, these hexagonal plates become unstable because new water molecules attach more easily to the hexagon's corners, which stick out, than to their faces, so the corners grow into points, creating stellar forms. These in turn catch more water molecules in a positive feedback loop, and as these crystals grow larger, they branch and become more complex still, turning into stellar dendrites.

A snow crystal is therefore the product of its temperature and moisture history. As they begin to form in the clouds, swirl around and fall through the sky, they are subject to a complex range of conditions, which makes for complex crystals. Their symmetry derives from the fact that one side of a crystal will often have a similar history to the other side, and because each crystal's history is different, snow crystals are all different. The fact that the snow crystals carry so much information about the conditions that form them led Nakaya to call them 'letters from the sky'.

Even though scientists can now decipher these letters, we have yet to fully master the language of snow. The physicist who today wears the mantle of Bentley and Nakaya is a CalTech professor named Kenneth Libbrecht. Libbrecht admits there is still much to be learnt about the growth of snow crystals. Why, for instance, do snow crystals grow into plate-like or columnar forms, and why do they switch from one to the other with temperature? He has been working on this for some years now, but the answer, he says, remains elusive. 'It seems like a simple enough problem, until you dig into the details. The large-scale structure of a snow crystal is determined by the detailed molecular dynamics at the ice surface, which we do not understand very well.'

In his biography of water, Philip Ball writes: 'No other substance is known to crystallize in quite so many different ways. Our most familiar image of a snowflake is that of a flat star – and why the ice crystal should confine itself to a plane, when all three dimensions are available, is still something of a mystery.'

Looking again at Bentley's photographs, it is easy to see why, in successive ages of theologians, philosophers and scientists, the apparently unnecessary beauty of snow crystals has been put forward as evidence of a divinity. Kepler believed the design, like the structure of the universe, was God's. Water vapour had a soul, he wrote, and the soul of water created snow crystals. 'Thus in vapour too, which as a whole had been possessed by a soul, there is nothing to wonder at, if, when cold engineers the break-up of the uniform whole, the soul should be busy with forming the parts.' In *Micrographia*, Robert Hooke ranks snowflakes with other symbols of holiness: 'I could very easily, and I think truly, deduce the cause of the curious six-angular figures of snow, and the appearance of haloes etc.' William Scoresby Jr, who left whaling to become a minister of the Church, wrote that the 'particular and endless modifications' of snow crystals 'can only be referred to the will and pleasure of the Great First Cause, whose works, even the most minute and evanescent, and in regions the most remote from human observation, are altogether admirable'.

In Jericho, Ray told me that Bentley was no more religious than anyone else at the time, and yet his writing about his beloved crystals invokes the spiritual. In an article published in the *Christian Herald* in 1904 he wrote: 'The snow crystals... come to us not only to reveal the wondrous beauty of the minute in Nature, but to teach us that all earthly beauty is transient and must soon fade away. But... like the beauties of the autumn, as of the evening sky, it fades to come again.'

Beauty fades, people die, civilizations rise and fall. Our lives, like those of snow crystals, are transient moments within a succession of other transiences.

We resemble those crystals in another way, too. Like them, we are made mostly of water. When we die, the water in us will find its way to the sea, where in time it will be lifted up by the sun, to fall again as snow.

CHAPTER 4

Snow Falls on Gotham

Syracuse, Buffalo, New York

Of the countless winds and storm systems that blow around the Earth, some have become famous outside the professional interests of meteorologists, farmers and sailors. Among these are the Mistral, which brings low temperatures to the south of France; and the Chinook or 'snow-eater', known in the Alps as the föhn, which accelerates downhill from the tops of mountains and whose dry warmth can clear a field of ice crystals a metre deep in a few hours. On the West Coast of the United States they talk about the Pineapple Express, which transports moisture from Hawaii into California, Oregon and Washington, while in the Midwest the snow is sometimes delivered by Panhandle Hookers, which originate in Texas. Two of the principal systems of the US East Coast are the Alberta Clipper – with its colourful variants, the Manitoban Mauler and the Saskatchewan Screamer – and the Nor'easter, which is formed when an area of warm, moist air generated by the Gulf Stream collides with cold air coming down from the Arctic. This great meeting of air masses can produce a vast area of low pressure that moves up the eastern seaboard, rotating anticlockwise and bringing north-easterly winds to the coast from New Jersey all the way to Quebec. Depending on the warmth of the Atlantic and the coldness of the northerly airstream, Nor'easters can produce light airs and gentle rainfall or hurricane force winds and snowdrifts twenty feet deep.

The distinctive types of East Coast system have occasionally produced individual storms so devastating that they have also become household names. So the infamous Groundhog Day Gale of 1976 was actually a vicious Saskatchewan Screamer, and a particularly savage nor'easter was dubbed the 'Perfect Storm' by the writer Sebastian Junger after it sank the Massachusetts fishing vessel *Andrea Gail* in 1991. A nor'easter was also responsible for the most notorious snowstorm in American history, the Blizzard of 1888, which brought late-nineteenth-century New York to a standstill.

I had no idea that my very own nor'easter was on its way as I drove back to New York City from Vermont to pick up Lucy and the boys. Nor was I aware that the family holiday we had planned in early spring of 2006 – mixed with a sprinkling of snow tourism – would be quite so inundated with ice crystals. But a light air was stirring further south which in coming days would drop a small quantity of snow on the Appalachian Mountains and move out into the Atlantic, where the unusually warm February sea would stoke it into a Category 3 event on the National Weather Service's snow impact assessment scale, which indicates a 'major' system. It would then dump the most snow ever measured on the New York weather station in Central Park. However, as I pulled off the New

Jersey turnpike towards Newark, everyone, even the meteorologists, was oblivious to the coming storm.

It was a treat to see the boys run down the walkway into the arrivals hall. They were different in many almost imperceptible ways since the last time I'd seen them, and in one very obvious one: they were dressed as Batman and Superman. They were in the grip of a superhero craze then. They went everywhere in their pyjamas, which they wore under their clothes so that they could change instantly whenever heroes were needed. They asked me about the Arctic. Were there penguins or bears? Lucy waited a while before she told me her news. She was pregnant. We drank a toast with the only liquid to hand, Diet Coke.

After the flight from London they were all tired, but there was no time to lose. We had a couple of hundred miles to cover before bed. I wanted to get to Syracuse in central New York State to observe a special weather phenomenon: the 'lake effect', which is produced by a mini climate system surrounding the Great Lakes and makes for some of the heaviest snowfalls and best crystals in the world.

We drove away from the coast through the heavy industrial landscape of New Jersey, where giant steel constructions march like dinosaurs into the west while the giddy towers of Manhattan retreated in the rear-view mirror. The Delaware River was in flood when we crossed it at Water Gap, and we continued north through the low hills of Pennsylvania along an interstate lined with thickets of dark wood that in late spring would have been covered with coloured leaves. They were all asleep by the time we reached Scranton, a hundred miles from New York City. It now began to snow lightly from the iron clouds which obscured the late-afternoon sun. Before long the air was white and the trees invisible. Snowflakes were steadily striking the windscreen in lumpy splats before the wiper blades swept them aside into a growing wet drift. The tyres hissed through the crystals and the steering wheel began to pull towards the edge of the road. Trucks sped by regardless, sloshing bow waves of snow and rocking the car with their wakes. I saw a saloon that had spun off and ended up in the wide central reservation and was grateful, for once, to be inside the massive Chevrolet.

As dusk turned to night I switched on the main beams, but the reflected brightness was blinding, so I dipped them and squinted into the furthest reaches of the ovals of light formed by the headlamps. The traffic markings had long since been covered over and for a time, two dark lines of tyre tracks in the right-hand lane marked where the road was, but soon these were lost in the thickening accumulation of snow. On a long straight, I put my foot down to catch the dim red tail-lights of the pick-up in front, keeping it in sight as the car behind kept hold of us, and in the company of these two strange vehicles I navigated my sleeping cargo the last fifty miles until we reached the safety of the streetlamps of Syracuse. Some time after midnight I found our hotel on the edge of town. It was basic but warm and clean, with an open fire in the lobby. I carried Harry and Arthur inside in their pyjamas.

It was snowing gently the next day as we walked out to the car park with the sun shining weakly through cloud. The tarmac was covered with a crystal fuzz that came up to my shins – airy, dry snow which made a wispy noise as we walked. I lowered my face to the car's voluminous bonnet and looked across its glittering surface. Other than in Bentley's photographs, I had never seen such perfection: a hundred thousand branching stars, each catching the light at its own angle, and each angled face reflecting a different shade of white. I put my hand into the layer, which was three inches deep, and came away with a whole clump. I could clearly see the roots and branches of the perfect stellar dendrites. Each hexagonal form, the crossing of three diameters, tangled with the spines of its neighbours so that when I scooped up a few hundred on my fingertip they stayed there, holding the shape they had landed in, a soufflé of frozen cloud. More of this soufflé lay around the parking lot in big heaps. Soon much of it would collapse under the revving snowplough, a four-wheel-drive with a blade at the front which was hurrying to clear a route to the main road. But even when the plough had done its work there would be millions of perfect crystals left, glittering on either side of the cleared path.

Eventually I had to push the dust off the windscreen and windows so that I could see out. As the sun brightened others began to scrape and de-ice their vehicles around us. Drivers talked and smiled at each other and made jokes and wished each other good morning. Is this kind of snow normal around here? I asked the driver nearest me as she scraped away. 'Sure,' she said. 'It's the lake effect. It's a winter wonderland!' Another neighbour, catching sight of my son Arthur rubbing at a patch of snow on the bumper with his glove, produced a device four feet long with plastic planes and spikes sticking out at all angles. 'Here you go, try that one,' he said.

Frank Capra is thought to have based the fictional town in *It's a Wonderful*

Life on Seneca Falls, a few miles down the road from Syracuse. It was almost as if we had stepped into a scene from it.

'It's like a fire station here,' said Tom, the meteorologist in charge at the National Weather Service bureau at Buffalo, the following day. 'Generally there's not much happening but when severe weather comes in the adrenalin really starts to pump.' I had come to hear how the lake effect worked from the world expert on the subject. I also wanted to know what it was like to live in a city with so high a snowfall. How did Buffalo, Syracuse, Rochester and the towns in between deal with it? We had driven that morning along the southern edge of Lake Ontario, moving in and out of fingers of clouds as we went, skipping from sunshine to snow and back again. I had dropped Lucy and the boys in a hotel with a swimming pool – I would only be a couple of hours, I said – while I drove to meet Tom at his low, brick office near the airport.

He spoke rapidly as we walked past the forecasters at their bank of desks, each of which was equipped with a number of computer screens, and ushered me into a small lecture room with several chairs and a screen for giving Power-Point presentations. 'People here work on storms to see if there are any signs that they are going to turn severe,' he said. 'Our warnings can save lives. We know that, because people ring in to tell us they heard the warning on the radio and took cover, and if they hadn't they wouldn't have made it.'

Tom had grown up in Buffalo. As a meteorologist he was a veteran of the extreme events that occur in this weather nexus on the lee shores of both Lake

Ontario and Lake Erie. Buffalo's most severe snow event was a blizzard which struck the city in January 1977. The trouble that year was that there was already thirty inches of snow before the wind came along, Tom explained, and the snow hadn't been through the freeze-thaw process, so it was just lying there ready to be blown around, fuel for the blizzard. Then the wind came and blew incessantly for five days. Along with the arctic temperatures this produced a life-threatening wind-chill as well as reducing visibility to a few hundred feet.

'Individuals in their cars got stranded, as the snow drifted across the roadways,' said Tom. 'Visibility was so low they couldn't leave their vehicles. They either froze to death or were poisoned by carbon monoxide from keeping the engine running as the snow blocked up the exhaust and filled the car with fumes.' At a conservative estimate, the Blizzard of 1977 killed twenty-nine people.

Almost twenty-five years later, at about eleven o'clock one Monday morning in November 2000, another storm hit Buffalo. This was the one the National Weather Service named Chestnut. Chestnut showed that the city hadn't learnt the lessons of the Blizzard of 1977. A band of cloud fifteen miles wide began to drop three inches of snow an hour on Buffalo. Employers and schools immediately started trying to get people home, Tom said, which had the effect of flushing out all the cars on to the major roads. However, a couple of trucks had jack-knifed on the ramps to the interstates and blocked them, so the vehicles that were trying to get people home couldn't move, and once the trucks had been moved off the ramps the rest of the traffic was snowed in because the flakes were coming down so fast. There was gridlock. Thousands of people were caught in their cars, and the pressure on the cellphone network meant it broke down, so no one could call their friends and families to tell them where they were. Only two things saved Buffalo from another major catastrophe during Chestnut: the temperature, which wasn't too low, and the wind, which wasn't too strong.

Why did the lakes produce such extraordinary snowstorms? I asked.

Tom, who was animated now, began his explanation from first principles. 'Imagine a warm summer day,' he said. 'The sun warms the ground and there's a little bubble of air that gets warmer than the cooler air around it. When air heats up it gets lighter and rises. At some point the bubble of air will start to cool, and the moisture within condenses out as cloud droplets. If you have enough of these packets of air, they create a cloud.' Inside these clouds, snow crystals form, seeded by small specks of dust or pollen, and grow until they become heavy enough to fall to earth.

The lake effect took place when cold, treacly Arctic air moved down from the north and crossed the relatively warm body of water in the Great Lakes. 'In the lake situation,' said Tom, 'the little air parcels over our warm water start to warm up too. They say, hey, we're getting a little buoyant here, and start rising, and moisture condenses out.'

Because the supply of warmth and moisture from the lake is static, as long as cold air keeps moving across it, clouds will continue to form. These conditions can work steadily for days at a time, producing a constant stream of cloud and enormous volumes of snow that often fall on a very small area which is tightly governed by the bearing of the wind. If the wind shifts a few points, the snow will shift too. It was like sticking one end of a hosepipe into the lake, with the wind deciding which direction to point the other end.

The effect of the big snowstorms on the city depended on the day of the week they hit, and even on the time of day. Six inches on Saturday night would cause very little problem in Buffalo as everyone stayed at home, but two inches in a rush hour could be lethal. Chestnut was dangerous because it hit on Monday morning, and even though it didn't directly kill anyone, the National Guard had to be called out, the airport was closed for twenty-four hours and $43 million worth of damage was done.

What was it like living with so much snow? 'When I travel to conferences and say I'm from Buffalo they say, "Oh, you get so much snow," like it's a bad thing,' said Tom. 'But I look at it the other way round. There's also a tendency for people to have a self-deprecating sense of humour, like, "We're tough up here, we can take it."'

The ability to laugh at the snow and get on regardless reminded me of something Rasmussen had written about the Inuit, that in their extreme climate they thought it was important to be happy: 'A happy person was considered a capable person, a reliable person – in short, a good person.' Tom, who loved winter sports, seemed extremely happy with his snow. I thanked him, and drove back to my family through a thin flurry.

As we travelled through central New York State over the following days, from the icy falls at Niagara in the west to Syracuse and the wooded Tug Hill plateau in the east, I discovered some of Tom's philosophical snow humour. One day on the plateau, which is criss-crossed with snow-machine trails, we drove through a tiny settlement called Fargo, which brought back to me the gentle comedy of the Gundersons, Lundegaards and Grimsruds depicted in the Coen brothers' film of the same name, and the way people said 'ya' all the time in the northern American way. *Fargo* was set in North Dakota, but when I bought a brand of bottled water that carried the line 'Snowflakes Never Tasted So Good', I thought it could have been written by the Coens.

As Tom had said, central New Yorkers had a way of poking fun at their climate. Typical jokes run like this: You've lived in Syracuse/Buffalo (the two cities are interchangeable for these purposes) too long when... 'You have more miles on your snow-blower than your car'; 'You think driving is better in the winter because the potholes get filled with snow'; 'Your grandparents drive at 65mph through thirteen feet of snow during a raging blizzard without flinching.'

Other jokes are more subtle. Syracuse Newspapers once ran a tongue-in-cheek Snow Edition, offering, among other things, features on the largest snowplough in the world and advice on how to shovel snow ('Push. Don't lift.' 'Take breaks and stretch.' 'Don't do the bend and twist.' 'Don't force it.'). Perhaps the most elaborate example of New York snow humour is the competition that takes place every year between the cities of Buffalo, Syracuse, Rochester, Binghampton and Albany, called the Golden Snowball. This vaunted trophy – originally it was a tennis ball spray-painted gold but now it is a snow-globe – is awarded in spring amid great pomp to the city that has received the most snowfall over the previous winter. At the peak of the Golden Snowball's popularity, in the 1980s, a thousand people came out to watch the presentation, bands played and representatives of snowbound places were invited from all over the world. Local newspapers still carry tallies of the daily aggregated snowfall of each city on their back page.

If there is a serious rivalry between the cities over snow it lies in their ability to clean up and carry on after a storm, which is something at which central New Yorkers excel. Snowfalls that would immobilize other places for a week cause barely a ripple here.

As we edged out into the traffic after de-icing the car that first morning, I was struck by the fact that every tenth vehicle I counted was a privateer snow-plough, a four-wheel-drive pickup with a hydraulically operated blade at the front which cleared car parks and driveways. These were in addition to the municipal snowploughs that were piling around the roads.

Inevitably, there are those who hate the snow. A pair of professors at Syracuse University told me that their maintenance staff boast that snow will never close the campus. This surprised colleagues arriving from the south. One new arrival, they recalled, had lasted only until her first snowstorm. The following morning, she hadn't shown up for work because of the snow, although everyone else did, as normal. A few days later, she handed in her notice and caught a plane back south.

Inside our vast Chevrolet, an area of low pressure was developing. The days of suite hotels on the outskirts of rust-belt towns had taken their toll as we drove round this industrial winter landscape, staying on one ring-road after another and eating in neon-lit diners. The jet-lagged boys woke up in the night, disoriented, or very early in the morning with manic, overtired grins. This was no one's idea of a good time, particularly not Lucy's.

The superhero fad had narrowed to the one DVD they had brought from England, a collection of *Spiderman* episodes from an animated series in the 1970s. They refused to watch anything else, and although it kept them occupied as we travelled along the freeways at 60mph, the theme tune, repeated over and over as Doc Ock fought Spiderman in a variety of New York warehouses, began to get to us.

Lucy is a heat-lover, happiest on the beach. Here she felt permanently cold. Our family holiday had been sacrificed to my obsession. Soon the boys started to complain that they were uncomfortable in their padded trousers, thick coats and snow boots over their Batman and Spiderman suits, but they refused point blank to take them off. This was one concession they would not make.

As we sat in an eatery in a vast Syracuse mall complex, I was desperate to turn the mood. I had an idea: we should go sledging.

After lunch we crossed the empty tarmac to a shop with aisles as wide as roads extending across the building's many acres in every direction. The only people in the store that weekday morning were its employees, who were either very young or very old, and the four of us. We wandered among racks of dresses and maternity wear, lingerie and shoes, accompanied by the heartfelt pleading of shopping rock, until, opposite a shelf containing a hundred styles of baseball cap, we found the sledges. They were all plastic, some consisting simply of a moulded seat and handle, some shaped like cut-down bob-sleds with steering wheels, others inflatable with camouflage patterns. We bought two and carried them back to the car.

An hour's drive out of town, we entered an area of white, rolling hills in a corner of Onondaga County that was latticed with snow-machine trails. We parked at the bottom of one of these slopes and climbed up it until the gradient was steep enough for our sledges to slide. The snow was dry and deep, and for the first few runs it was too thick for the boys to move without pushing, but when it was packed down and smoothed, two of us could climb on one sledge together and bowl down the hill until we stopped or fell off, laughing. After a while I became hot with the exertion of walking back up the hill and stopped at the top to sit and take in the view of the white farmland. The sound of the children's voices drifted up towards me.

From here I had a good view of the wooded hills and white-covered fields. Harry and Arthur were climbing up through the snow, red-cheeked and happy again. They reminded me of my brother and me as children, dragging our sledges up the hills of Yorkshire. I wondered if the boys would remember this childhood day in this faraway place. Perhaps many years later, something would stir their memory – the smell of snow or the feeling of it against their skin – and they would recall the time they spent together in the Adirondacks.

We left the sledging hill in the late afternoon and drove back to Syracuse to find somewhere to stay. We stopped at a gas station to fill up the Chevrolet, and when I got back behind the wheel Lucy showed me a copy of the newspaper she had found. As I read the headline, a prickling sensation began in my hair and worked down my back.

'A Record Snow,' it read. '26.9 Inches Fall in New York City.'

It went on: 'The biggest winter storm in New York City history – destined for lionization as the Blizzard of '06 – buried the region and much of the Northeast yesterday under blowing, drifting, thigh-high snows that crippled transportation and commerce, knocked out power and disrupted life for millions in 14 states.'

The whisper of wind that had begun further south had landed in New York City as one of the great storms of the East Coast, and while I was upstate investigating the lake effect, it had passed us by.

The nor'easter that had been stoking itself out over the Atlantic had become 'a great crab nebula' 1,200 miles long and 500 miles wide, according to the newspaper. Pressure had fallen so low at the centre of the storm that it had formed an eye, which is rare for a non-tropical storm. It had dropped great quantities of snow on the largest conurbation in North America, the heaviest downfall landing in a band that ran from Charleston, Maine, through Washington, DC, Baltimore and Boston.

It was clear that we had to get to New York City as quickly as we could. We could stay an extra night there before catching the flight home. Lucy was even keener than I was. We could make it that day if we drove hard.

As we raced through the evening towards the metropolis, listening out for updates on the radio, Lucy read out bits of the *New York Times* reports.

Snow had begun to fall on the city on Saturday afternoon. New Yorkers were woken by claps of thunder and lightning early on Sunday morning, and saw snow tumbling from the skies outside their windows. They had put on their boots, coats and hats, and had opened their doors to great drifts against the top steps and stepped gingerly out, inhaling the cold, watery scent, unused to their voices and footfalls in the sound-deadened streets. They had walked down the middle of car-free roads, examining the dustings on the trash-can lids and branches, and the buried cars, admiring the fresh white tablecloth that was now spread over the city, as they made their way into the gardens and parks and squares, dragging with them their sledges and inflatable tubes. Big dogs ran through the drifts. Cross-country skiers had been seen trekking across Times Square and down Park Avenue. Strangers had fallen into friendly conversations.

On Sunday afternoon, after the storm had abated, the security guards who operate the weather station in Central Park Zoo measured a snowfall of 26.9 inches in twenty-four hours. Not since records had begun in 1869 had so much fallen in a day.

There had been disruption and delay, including power outages in New Jersey and Long Island, which left thousands without electricity. The snow around Central Park had been so deep that a plough got stuck, and the plough that was sent to rescue it got stuck too. For the first time since 11 September 2001, JFK, La Guardia and Newark airports had been closed, as was Ronald Reagan National Airport near Washington. Two thousand five hundred flights were cancelled in the New York City area alone. On Sunday night, a Turkish Airways flight had slid off a runway at JFK. Snow had drifted across the tracks of the Long Island Rail Road, which meant the trains stalled. Tracks were cleared with snowplough and icebreaker trains, or trains with jet engines attached to them that were used to melt snow in the marshalling yards, or by men and women armed with brushes and antifreeze.

It was 'a remarkable and relentless fall' that had 'eclipsed the legendary blow of Dec. 26–27, 1947, which dropped 26.4 inches and killed 77 people. It also easily surpassed the memorable No. 3 and No. 2 storms, of Jan. 6–7, 1996, which left 20.2 inches, and March 12–14, 1888, the notorious Blizzard of '88, which dropped 21 inches.'

My God, it was even bigger than '88.

The nor'easter that engulfed the East Coast in March 1888 gained such notoriety that it has garnered several different names: the Great White Hurricane, the Blizzard of '88, and sometimes simply the Great Blizzard. Although the last survivors are long dead, the storm's effect on New York and its place in the city's collective memory has lived on. The Blizzard of 1888 demonstrated to a society growing ever more confident of its technological power that civilization could still be brought down by the weather.

The storm hit New York in the early hours of Monday 12 March, when the rain that had been falling on the city began to turn to snow and the wind started to build. By 7 a.m. the New York Weather Station reported 10 inches of snow on the ground. By 3 p.m. there were 15.5 inches, and it had been blown by 50-mph winds into drifts twenty or thirty feet deep. Yet the work ethic of the time was strong enough, and employees were so concerned about job security, that, faced with doorways and stoops blocked by snow that Monday morning, people jumped out of windows and into drifts in order to begin their journeys to work. Many of them would not get to their destinations that day, or even the next, and some would never get there at all.

New York's mass transport system in 1888 was made up of three types of conveyance. There were the elevated railways, or Els, which were pulled by scaled-down steam engines; the horse-drawn buses, or 'horse cars'; and the railways that brought commuters into New York's famous stations. In the event of a great snowfall, without a reliable or effective way to clear the snow, the El tracks soon became blocked and the telegraph that kept El stations in contact with the network controllers broke down – as did telegraph wires and power lines right across the Northeast – so there was no way for trains to communicate: those that were able to move crashed into others that were stationary. Many simply got stuck, and had to wait for help that didn't come, leaving their passengers stranded. Faced with the prospect of being trapped for hours in a freezing carriage, people decided to crawl down the elevated tracks in the blizzard to try to reach the next station. Others were rescued by enterprising businessmen who put ladders up against the track and charged people to come down.

On the outskirts of the city, the railroads leading into the New York terminuses were becoming blocked, and a hush descended even on the habitual cacophony of Grand Central. On the Monday, a reporter for the *New York Sun* visited the station, which was usually the busiest on the continent: 'The interior... and the yard immediately outside, where the hissing of steam and the clanging of bells is heard day and night, were as silent as the grave,' he wrote. 'Clouds of snow were driven into the immense station, whirled up to the transparent ceiling, and again fell softly on the long lines of trains... no sound was audible in the great structure save that of the moaning of the wind.'

Hundreds of trains had become stuck in snow on the lines into the city. Some of the drifts were deeper than the engines were high, and commuters were trapped for several days until the storm had blown itself out. During their entombment, passengers worked their way through the carriage stove fuel, then started to chop up the seats and tables of the cars, as the blizzard whistled through the gaps in the coachwork. The people who had brought packed lunches and snack food shared them out among the other passengers, and when these were gone, the strongest were despatched to farms and houses in the surrounding countryside to beg or forage for something for the snowbound commuters to eat. Those who abandoned the trains to walk home found themselves struggling for hours through drifts up to their armpits, and were often suffering from frostbite by the time they reached their houses.

In the city centre, meanwhile, the drivers of the horse-cars that set out on Monday morning soon found the winds and drifted snow impassable, even with horses double- or triple-teamed. Many horse teams were abandoned, which only exacerbated the blockages in the streets. Those who escaped the Els and horse-cars, or avoided them altogether and tried to make their journey through the city on foot, came across horses that had frozen solid in their harnesses and whose heads stuck up out of the drifted snow. The wind was so strong that unlucky pedestrians were blown bodily into the drifts and found they could not

get out. Women suffered more than men as their billowing dresses, hats and

high heels made it harder for them to remain on their feet. Witnesses reported

women being blown over repeatedly, or simply falling on the icy pavements.

The bodies of men and women who had been pushed by the wind into drifts

were discovered dead hours or even days later by an arm or a leg protruding

stiffly from the snow. One wealthy businessman, George Baremore, walking

in the relative calm beneath the El tracks, felt soft thumps on his head, and real-

ized that frozen sparrows were dropping on him from their nests in the trestles

above. Sometime later he was found dead in a snowdrift. Caught by the wind,

he had struck his head against an El support.

As the storm continued into Monday night, New York's ability to keep

itself warm, fed and healthy deteriorated. The milkmen couldn't deliver, so

children and babies went hungry. Coal ran short as delivery carts couldn't get

through, and those that became stuck were instantly surrounded by people

pressing to buy whatever was left. Such was the hunger that some were reported

to have eaten the dead sparrows that had frozen in their thousands in trees across

the city. There were no mobs or looters, but many people were quick to turn the

storm to financial advantage: prices for essential commodities shot through

the roof. Even newspapers, which rushed out special Blizzard Editions, were

selling at $10 a copy.

By Tuesday morning the storm had eased a little, though snow was still

falling in the afternoon and temperatures remained perilously low and wind

speeds high. By Wednesday, temperatures finally rose above freezing. In all, fifty inches of snow had fallen. The drifts remained for many days afterwards, melting and causing floods, as gangs of shovellers made up of newly arrived Italian immigrants, some straight off the boat from Naples, cleared the snow into the rivers. At the end of the week, 400 people had been killed, 198 ships had been sunk or damaged in and around New York harbour and 800 bodies were waiting to be buried in the city cemeteries.

The most famous victim of the storm was Roscoe Conkling, a former Republican senator for New York and crony of the former President, Ulysses S. Grant. At the age of fifty-eight, Conkling was still a handsome, strapping man with a barrel chest who was abstemious with alcohol, didn't smoke and worked out daily with barbells and punchbags. He was intelligent, a wonderful orator and attractive to women. He was also vain and arrogant. According to one of his enemies, Rutherford B. Hayes, who had succeeded Grant as President, he was obsessed with 'the monomania of his own importance'. Chancey Depew, a contemporary, wrote of Conkling that 'his intolerable egotism deprived him of the vision necessary for supreme leadership'. If late-nineteenth-century New York's tendency to over-confidence could have been embodied in one man, no one was better qualified than Roscoe Conkling.

On the afternoon of Monday 12 March, 1888, when the New York blizzard was in full spate, Conkling had left his office in downtown Manhattan to return to the hotel in Madison Square where he had rooms. A young lawyer named William Sulzer saw him leave and offered to share a cab, but when Conkling heard that cab drivers were demanding up to $50 he said: 'I don't know about you, young man, but I'm strong enough to walk' and set off through the snow at a cracking pace with Sulzer struggling to keep up. The two men fought their way for a few blocks to the fashionable Astor House hotel, where Sulzer decided to give up. Conkling, a central New Yorker who was not going to be faced down by anything so trivial as a snowstorm, continued on. It was a brave but unwise decision.

'I went magnificently along, shouldering through drifts and headed for the north,' he told reporters the following day. 'I was pretty well exhausted when I got to Union Square, and, wiping the snow from my eyes, tried to make out the triangles [pathways that criss-crossed the park] there. But it was impossible. There was no light, and I plunged right through on as straight a line as I could. Sometimes I have run across passages in novels of great adventures in snow-storms; for example, in stories of Russian life where there would be a vivid description of a man's struggle on a snow-swept and windy plain; I have always considered them exaggerations, but I shall never say so again. As a matter of fact, the strongest description would fail to approximate the truth.'

It took Conkling three hours to reach Madison Square. There, attempting a short-cut, he became stuck in a huge snowdrift. 'I had got to the middle of the park and was up to my arms in a drift,' he said. 'I pulled the ice and snow from my eyes and held my hands up there till everything was melted off so that I might see; but it was too dark and the snow too blinding.' After twenty minutes he had freed himself from the snowdrift but had come 'as near giving right up and sinking down there to die as a man can and not do it. Somehow I got out and made my way along.' He finally reached the New York Club on 25th Street, 'covered all over with ice and packed snow', and collapsed on the doorstep. He was carried from there to his hotel but by then was extremely ill with mastoiditis and pneumonia. He died a month later.

The newspapers blamed the city's devastation on a similar mindset to that of Conkling. Before the storm, New York had been high on its own success. Bolstered by waves of immigration, on the brink of an extraordinary building boom, with advances in infrastructure and engineering that included the tele-phone and electricity for lighting, it was on its way to becoming not only the pre-eminent American city but the capital of the world. But although the metropolis was expanding rapidly, it was becoming increasingly dependent on technology to keep it moving. The elevated roads and the trains simply hadn't been designed to work in a severe blizzard and – perhaps the greatest sin of vanity – people had assumed that they would. The *Hartford Courant* ran an editorial that captured the public mood: 'It is the boasting and progressive 19th century that is paralyzed, while the slow-going 18th would have taken such an experience without a ruffle... There comes a snowstorm... there is no railroad, no telegraph, no horse-car, no milk, no delivery of food at the door. We starve in the midst of plenty... It is only a snowstorm, but it has downed us.'

In the storm's aftermath a tremendous public debate began about how to prevent the city from being so vulnerable again. One of the most significant long-term consequences for New York was the subway system. Just weeks before the blizzard struck, Mayor Abram Hewitt had introduced a bill outlining plans for an underground railway. The bill had been rejected, but now every New York paper was clamouring for more reliable and weather-proof trans-port. On the Monday after the blizzard, the *New York Times* produced a detailed blueprint for a subway that would run the length of Manhattan from Battery Point: electrified trains with electric lighting running through tunnels ventilated by shafts that ran up to the surface. This scheme would be adopted, more or less intact, fifteen years later.

The telegraph system benefited from the storm too, as it was decreed that the communications blackout between the Northeastern cities and the rest of the world must not happen again. Cables and power lines which had been carried high above the streets on great telegraph poles, and which had been swiftly brought down by the storm, would in future be buried underground.

But while chaos reigned in New York, the city's residents were showing a resilient sense of humour about the snow, and even a fondness for it. The *Morning Journal* on Tuesday 13 March published a Blizzard Extra 'Icicle Edition' and Special Snow Sheet. Mary Cable, who has written the most extensive account of the blizzard, relates how humorous slogans began to appear outside shops and banks, written either by the proprietors or by passing wits: 'It is better to see the rain rain than the snow reign', 'This way to the Klondike', 'Don't touch this snow, it's all we've got' and 'Snow for sale! Come early and avoid the rush!' Decades afterwards, people were still commemorating the blizzard as a special time when the people of the city had been brought together. In 1929, a group of veterans of 1888 started a society whose purpose was to gather survivors together to share experiences at an annual dinner. Members of the Blizzard Men and Blizzard Ladies remembered the storm as 'a great communal adventure'. 'It had its tragic side but curiously left in its wake mainly good will,' one of them said. 'To its survivors it has become almost a household symbol, standing not only for the storm itself but also for all that was best in the "good old days".'

On Monday 12 March, 1888, the most spectacular day of the New York blizzard, Wilson Bentley, the Snowflake Man, was in his woodshed photographing a grand total of twelve crystals, the largest number he had ever taken in a day. They were, he wrote, 'all splendid'.

It was dark when we parked outside the hotel on Amsterdam Avenue and wheeled our cases along the snow-covered sidewalk. The lobby floor was wet with melting snow that people had trodden in from the street. Though snow still lay deep on the sidewalks where it had been shovelled into stacks, the weather was strangely warm, and the streets were dripping.

The following morning we sweated in our winter coats as we walked towards Central Park. A man was trying to dig his car out of the drift that had almost covered it, but most people seemed to have left their vehicles to thaw out on their own. In the park, the snow was still a foot deep, and the boys sank into it up to their knees, fell over and threw snowballs at each other. Street sweepers were still clearing the footpaths. The playground was locked up because the snow was too deep.

In the south-east corner of Madison Square Park a life-size bronze stood under an inelegant mantle of melting snow. The sculptor John Quincy Adams Ward had set the handsome Roscoe Conkling in statesmanlike pose, his right foot forward and right arm cocked in the manner befitting a New York senator

and one-time US presidential candidate. The statue was erected in 1893, after
Conkling's friends petitioned the mayor and the park board. Now he stood
in the square where he had encountered that fatal drift, looking out across
the snow.

Since 1888, the careers of several American politicians have crashed or
soared on the basis of how they deal with the problem of snow. When a storm
blanketed the city in February 1969, New York mayor John V. Lindsay came
close to losing his job after the residents and councillors of Queens accused
him of bodging the clear-up. Chicago mayor Michael Bilandic lost his job in
1979 largely because of his slow response to a heavy snowfall, and Washington
mayor Marion S. Barry was called before a senate sub-committee for his sluggish reaction to a snowfall in 1987. Michael Bloomberg, New York's mayor at
the time of the 2006 blizzard, didn't make this mistake. He called a press conference so he could be seen on television standing shoulder to shoulder with his
sanitation commissioner, John J. Doherty.

Because of the warm winter, Commissioner Doherty had been able to
set about the clean-up operation with most of his $31 million annual snow-
clearance budget intact. Two thousand five hundred people had operated
snow-clearing equipment on twelve-hour, back-to-back shifts. Three hundred
and fifty salt spreaders had been deployed, which would drop more than 50,000
tonnes of salt, and more than 2,000 snowploughs had worked to clear the
streets. The city had recently bought twenty snow-melting machines, nick-
named 'hot tubs', for $200,000 each, which could turn 60 tonnes of snow an
hour into water before dropping it into the sewers. Two thousand people had

been hired at $10 an hour as emergency snow clearers, helping to shovel bus stops, street corners and drains. Freelance shovellers got in on the work too. Some were teenagers trying to make a bit of extra pocket money while others were poor and unemployed people from housing projects who travelled in to the smarter suburbs to make a few dollars whenever there was a snowfall. One 51-year-old man who, with his nephew, was shovelling snow in Brooklyn told the *New York Times*: 'If I don't do it, I can't eat. I don't have a regular job.'

No one in New York was killed by the Blizzard of '06. Some lives might even have been saved, Police Commissioner Raymond Kelly observed to one journalist, because the snow made people nicer to each other. Snow had even brought the murder rate down: from Saturday afternoon to Sunday evening, no homicides were reported compared with six during the same time the week before.

The city had been well prepared. The storm struck at the weekend. The temperature was not low and the winds were not strong. But the next time a dangerous storm comes along, with higher wind speeds and lower temperatures, during rush hour, will the city escape so lightly? Probably not. Although we collect vast quantities of data and analyse the results, the weather can still take our technology-dependent societies by surprise. Forecasters can still get it wrong, and even when they get it right people don't always listen. Councils and governments are still reluctant to spend public money planning for what are short-term possibilities, even when they are long-term certainties.

I couldn't help trying to imagine what Billy's ancestors on Baffin Island would have made of the Blizzard of '88. Perhaps they would have wondered why these southerners felt that life should go on as normal even when the weather turns foul. They might have reckoned it was a measure of how detached the white man was from the natural world.

That evening we went out into the neon streets to find something to eat. It was late, but Lucy and I had suddenly realized it was Valentine's night.

We went into the first restaurant we found, a chic-looking Italian. Bedraggled, with two of us in pyjamas, we approached the *maitre d*'s lectern anticipating a brusque 'I'm sorry, sir, we're full'. But they weren't full, and we were led through into the darkened dining room. It was packed with couples and raucous with laughter and shouted conversation. Storms create a sense of urgency in the living. In March 1888, Mary Cable noted that the bars of New York were packed and noisy, as they were in Buffalo in January 1977.

The superheroes were fussed over by the waiters. Our fellow diners pronounced them 'adorable' and 'wonderful'.

Later we tucked them up, the Man of Steel and the Dark Knight, in the big bed.

CHAPTER 5

Playing with Gravity

Chamonix, Argentière, Wildstrubel

It is hard to exaggerate the effect of the first big mountain on a lowlander. One hazy summer morning when I was twelve I remember walking along the Jungfraustrasse in Interlaken and thinking the pink and mauve peaks that loomed over the town were unusual clouds. My mother told me what they really were, but I didn't entirely believe her and kept looking up to check whether they had blown away. The next day we climbed into the mountains. We caught the rack-and-pinion railway to Grindelwald, which sits beneath the white wall of the Jungfrau range, and gawped at the towering black triangle that is the north face of the Eiger. From Kleine Scheidegg the train carried us up the inside of the mountain to the highest station in Europe, the Jungfraujoch, where we walked out on to the glittering wet summer snow to look at the view we had witnessed the previous day from the other side. It was a clear afternoon and the lakes and towns and mountains were tiny and distant, but it was the emptiness between these two worlds that impressed me most, the great vacuum which

I felt the urge to leap out into, like Icarus, and to hell with the consequences. Thinking about it even now gives me a tingling sensation in my wrists.

A similar feeling appears to have gripped John Dennis, a young Englishman who crossed the Alps in the summer of 1688 and described his experiences in a letter to a friend who had never seen mountains. 'We walk'd upon the very brink, in a literal sense, of Destruction; one Stumble, and both Life and Carcass had been at once destroyed. The sense of all this produc'd different emotions in me, viz., a delightful Horrour, a terrible Joy, and at the same time that I was infinitely pleas'd, I trembled.'

Dennis had captured the mixed emotions of excitement and terror that Edmund Burke would convert into the doctrine of the Sublime half a century later, and which would launch a host of well-to-do tourists, artists, philosophers and poets into the mountains to experience the 'Horrour' and 'Joy' they inflicted upon the mind. Among those visiting the Lake District, the Highlands and the Alps in the following decades were Samuel Johnson and James Boswell, J. M. W. Turner, Coleridge, Wordsworth, Keats, Byron and Percy and Mary Shelley. The marvels of the natural world had never before come under such intense intellectual scrutiny.

Since that first overwhelming introduction to the Alps, I have travelled back as often as possible, principally for the snow and skiing, but also for the scenery: first to Obergurgl and Hochgurgl, then later to Salzburg, the Grisons, Savoie, Haute Savoie, the Val d'Aosta and the Dolomites. Each time I tried to push myself a little bit further: off-piste one year, heli-skiing the next, glacier-skiing after that. Awaiting me on our return from New York was an adventure

I had planned with my friend Nick which I knew would be our most difficult yet. We would travel the *haute route* that leads over the cols and glaciers from Chamonix in France to Zermatt in Switzerland, and we had hired Philippe, of the prestigious Compagnie des Guides de Chamonix, to look after us and show us the way.

Philippe was a short man, five foot nothing in his ski boots, with a tidy beard and a face which folded around his eyes along deep creases created by decades of exposure to wind and sun. In the spring light that flooded through the open doors of the Chamonix café where we met him, his long hair seemed to contain almost every shade available: it was in patches black, brown and blond. The most surprising thing about Philippe, though, was his hands. They were small, in keeping with his stature, and the well-kept nails gave them an air of neatness, but the knuckles were oversized, the flesh along the levers of his fingers full and taut, and at the base of his thumbs were large pads of muscle. As he wrote down the night stops of our planned journey on a scrap of paper, the heavy forearms that controlled the digits undulated powerfully. These were the hands of a professional climber; they had spent a lifetime hauling rope and hammering ice screws and holding on, or, as Philippe put it, playing with gravity.

The mountains had been nursed into Philippe. His mother had climbed with him on her back when he was a baby. At the age of ten, when he was growing up near the mountains in Albertville, he knew that he wanted to be a guide. In his teens he was making a name for himself as a downhill skier.

When we met Philippe, he was in his mid-forties and had a reputation for safety. During his years in the mountains he had lost several good friends, including one who was caught in an avalanche with him in Canada. The falling snow had knocked Philippe unconscious and buried him for an hour and a half before help arrived, and although he had broken several major bones he had been dug out alive. His friend had not been so lucky. Philippe was back on his skis the following year, but it had taken a further seven to get his head right, as he put it. Later, as the relationship between us began to deteriorate, I would wonder if he had ever really got it right.

As we set out early on our first morning on what Philippe said was to be a training day, I was just grateful to have him to instruct us. He had already been through our equipment, and on our backs we carried a range of tools for traversing snow: snowboards, snowshoes and crampons, along with telescopic ski poles, climbing harnesses, carabiners, shovels, avalanche probes and transceivers, water, food, and all the clothing we would need for a week in the high mountains. We hadn't been told where we were going, so we obediently followed our agile little leader in single file as he climbed up from the top of the Flégère ski area, our boots landing after his in the track he made in the crisp, early-morning snow. A short, steady, upwards traverse took us around the bottom of a sheet of rock. To our left the slope steepened sharply towards a ridge which scraped the cloudless blue sky. 'OK,' said Philippe. '*Bon*. So we go up there a bit.'

Philippe moved off uphill, kicking steps in the snow and testing each carefully before applying his weight. Nick followed, and then me, stepping into each footprint with as much care as I could muster, leaning in towards the snow-covered slope, digging in with the ski poles, trying not to look down, trying not to imagine the fall. Halfway up, the slope began to steepen further, and Philippe took out the rope and joined us together. The wind blew, striking the boards which stuck up like sails above our packs and threatening to pluck us off the mountainside. I leant in further still, my face at the level of Nick's boots, and climbed on. It did not occur to me that we were going right to the top until we were almost there, and then we were over the col, stumbling down out of the wind on the northern side of the ridge. Having skied in Chamonix for many years but without ever leaving the valley, we were now looking over an unfamiliar world, a new distribution of rocks and snow, white and black, and a view for many miles towards Geneva.

The straight descent from here would have meant a gentle, almost untracked swoop through deep snow, but we could not follow such a simple route. At the lower end, the valley was blocked by a steep-walled lake and there was no way out. Instead we must traverse around the eastern wall, keeping the highest line possible, digging our edges into the slope to keep the downhill slip to a minimum. '*Bon*,' said Philippe, and on we went.

The traverse was not as easy as it had seemed from the col. The snow was some of the stickiest and deepest I had encountered. Philippe, for some dark reason I did not yet understand, rapidly became agitated. '*Merde*,' he said, as I fell over for the fourth or fifth time and began the slow process of extracting myself. I had tried to keep my height – there were cliffs at the bottom of the slope, I had seen them from the col – but because I could not maintain speed I had been steadily progressing downwards. 'You must be efficient now,' Philippe said, in a voice clipped with anxiety, as he tramped back over towards me on his skis. 'Take off your board. Give it to me. Now walk uphill, on your knees.' The snow was so deep that when I put my boots in it I sank up to my waist, but moving on my knees seemed to spread the load. 'Now, put on your board.'

After a difficult half-hour we reached the end of the traverse, and were standing at the junction of a new valley which led back up towards the group of mountains we had crossed, but there was no let-up in our guide's mood. 'There is a danger of avalanches,' he said. 'Because it is warm we have little time. Eat something, drink something, but put your snowshoes on quickly. We must move.' Shaking after the physical exertion of digging myself and my board out of deep snow, I put on my snowshoes. I tried to drink, couldn't eat, then set out on the climb up towards the next col as if my life depended on it, which I believed it did. I was now moving too quickly for Philippe. 'You are walking like a crab,' he said, catching up, stopping me, and doing an imitation of my quickstep. 'No. Like this.' He measured out the steady rhythm that a mountaineer should use for ascent, 'One-two-three-four...' We set off again, Philippe in front, setting the pace. He was not fast but the effort of following a narrow ski track in snowshoes was exhausting. I stopped for a few seconds to catch my breath, and when I looked up he glared at me, and nodded towards the mountains to my right. He had seen what I hadn't: we were crossing an avalanche run-out zone, and he was looking up at the snow above us on the slope, calculating whether it was about to slide. I hurried forward again.

We reached the second col around noon. Dazed and breathing heavily, I took off my snowshoes.

From here, at least, it should have been a simple descent back to Chamonix. We looked north-west at the valley we had just ascended, a beautiful snow-filled trough which I knew now to be full of hidden danger. We sat on some tufts of grass near the col and talked about avalanches, about how, in the spring sun, the snow was pulling away from the rock and black glide-cracks were opening in the whiteness, about the fact that you don't hear an avalanche until it hits you. 'But if you are here,' said Philippe, gesturing at the nearby mountains, 'you accept the risk.'

I thought about this on the way down. When had I accepted the risk that day? I had not even known where we were going. Perhaps by 'here', he meant the mountains in general. By coming to the Alps, I had acknowledged that they might kill me. But in a small, cowardly part of my soul, I hadn't accepted that, didn't want it. Instead I wanted reassurance, a gold-plated guarantee that I was going home again.

Over the course of that first day, Philippe and I developed a mutual mistrust. We would never quite recover from it.

The mountains have long been a playground for those who can afford them. Now 60 to 80 million tourists visit the Alps each year, clocking up 160 million days of skiing in France, Austria, Switzerland and Germany. Chamonix alone hosts some 60,000 guest beds, and people come from all over the world to ski there. A large Alpine ski area like the Trois Vallées in France claims more than 600km of pistes and some two hundred lifts. Portes du Soleil, which links fourteen resorts in France and Switzerland, says it has 650km of pistes and around 230 ski-lifts. Even the smallest Alpine villages have a slow, clanking old chair or rope-tow, or runs for people to walk up. Skiing and its clutch of snow-sports cousins and offspring – telemarking, snowboarding, *Langlauf* – are part of the fabric of alpine society. More people visit the mountains in winter than in summer. For some it is a holiday; for some it is a living; for others a way of life.

Sitting in a restaurant on a mountainside sipping *Glühwein* while watching people snake down the mountain, skiing seems like a very modern phenom-enon, but as the pioneering British skier and travel agent Arnold Lunn noted in his *History of Ski-ing* of 1927, it is one of the oldest of sports. The ski histo-rian John Weinstock has written that skis and snowshoes were developed by hunters who needed to follow migrating herds north across the snow as the glacial ice retreated during the last ice age many thousands of years ago.

In the White Sea and Lake Onega areas of north-west Russia there are rock carvings believed to be four thousand years old that show men hunting on long skis, with holes in the snow beside their tracks where the huntsmen have been poling themselves along with their sticks (or in this case a spear and a bow). On the island of Rødøy in northern Norway, carvings from around the same time depict a man with tremendously long skis. The oldest physical evidence of skiing goes back even further: fragments of skis found in bogs in Russia have been carbon dated to 6,700 BC.

Skis and snowshoes were probably developed as a natural extension of Stone Age foot coverings, and one form of snowshoe might have been made of wood in an oblong shape. Although this shoe worked well, snow tended to stick to its sole, and so it was covered with animal hide. This not only prevented the build-up of snow, but the nap of the hide facing towards the rear stopped the skis from sliding backwards. Ski tourers still use this principle today, fitting synthetic 'skins' in order to climb.

The earliest documentary evidence of skiing is a Chinese book dating from between 211 BC and 206 BC. It describes the Dingling people who lived in the Altai Mountains of north-west China. The Dingling 'sped like goats in the valleys and on the flatlands, wearing the "horns of a goat" – a kind of knee-high fur boot under which is a wooden board with a hoof-shaped front tip'. Huan-jù kî, writing around AD 976 to 984, told of people in the mountains south-east of Kirghiz who, in hunting, 'make use of a foot covering called wooden horse. It is similar to a sled, but the head is high [i.e. curved upward]. The bottom surface is covered with horsehide so that the tips of the hairs run backwards. When the hunter has tied such boards on his feet and runs down a slope, then he overtakes the fleeing deer. If he runs over a plain covered with snow, he sticks a pole in the ground and runs like a ship; in this manner he also overtakes the fleeing deer. The same pole serves him as support when slopes are to be climbed.'

The skiing people most often mentioned in early Western writings are the people we now call the Sámi, or Lapps, but who were then known as the 'Scrithiphini', 'Scritobini', or, later, 'Skridfinns'. As the ancestors of the Sámi migrated from central Asia to their current homelands in northern Scandinavia, they took skiing with them. Saxo, the twelfth-century Danish historian, wrote

of them: 'The Finns have always travelled by gliding swiftly on smooth boards
and have complete control of their speed as they race along, so that men say
they can be there and gone in a flash, just as they please... the nimbleness of
their bodies and skis combined gives them a practised ease in attacking and
retreating.'

Archaeological finds show that ancient skis varied enormously. Each local
culture appears to have had its own skills and ideas and ranges of snow conditions for which different adaptations were best suited. Some skis were short and
wide, others long and thin, some had leashes attached to their tips. There were
also many different foot positions and types of binding, and degrees of taper
from the tip to the binding to the tail. A popular variation, which lasted well into
the nineteenth century, was for skis of different lengths: the longer ski would
be used for gliding and the shorter one for propulsion, by means of a scooting
push. The skis clearly gave the nomads a huge advantage, and the winter hunt
that was a struggle for a non-skier would have turned into a killing field as the
hoofed herds struggled through deep snow that the Sámi were able simply to
glide over. The animal populations were inevitably affected. By the thirteenth
century, laws had been enacted in Norway to protect elk in winter from the
hunter-skiers.

In Scandinavia, the Vikings learnt skiing from the Sámi, and for them it
became one of the essential skills of a warrior. Skis are featured in several Norse
myths, and there were even a god and goddess of skiing, Ull and Skrodi. King
Harald Fairhair, who ruled in Norway at the end of the ninth and beginning
of the tenth century, is described in the sagas as being with his court near a

mountain when he sees a man approaching. 'What is that,' he asks, 'that comes down the mountain like a whirlwind? Could that be a man on skis?' It was. According to legend, King Haakon (who ruled Norway between 1217 and 1263) was rescued from his enemies as a two-year-old by two skiers loyal to the reigning family.

In Sweden, the Vasa cross-country race celebrates the moment in 1522 when two skiers were despatched from Mora to catch up with the nationalist leader Gustav Vasa to tell him that the people were ready to join his war of liberation against the Danes. They met him at Salen, and the modern race covers the 85.8-km distance of the return journey to Mora. Peer Gynt, the folk hero on whom Henrik Ibsen based his 1867 play, was also an accomplished skier.

By the late nineteenth century, with the notable exception of a small skiing community in Slovenia, skiing appears to have existed only in Norway, Sweden, Finland and northern Russia. It was then that this method of snow travel moved out of the remote Norwegian hills and began its evolution into a modern sport.

During the night spent in a Chamonix hotel after our first experience of ski mountaineering, the confidence I had felt before we had started our trip began to slip away. This time, I had overreached myself. My habitual but manageable fear of heights was maturing into a fully-fledged phobia. The terror grew inside me so that when I closed my eyes I could only see myself plummeting through the snow into emptiness, as one rotten snow bridge after another collapsed beneath me, or as the blind ridge I was approaching fast gave way to a black precipice and I found it was too late to stop before I was tumbling through the air. My wrists tingled as I looked over the edge of some enormous drop, my fingers clutched at nothing. On imagined ascents, I felt my toes and heels bend to stop the wind sucking me backwards off the mountain.

For the first time I understood the terror that had led Alpine people to keep to the valleys and stay away from the snow. In the Middle Ages and afterwards, the mountains were thought to be full of malevolent spirits and strange beasts. I had read about the medieval belief that there was a ruined city, inhabited by the souls of the dead, at the top of the Matterhorn, and I was beginning to think it was plausible. The Matterhorn, at Zermatt, was my destination too, and in my current state of nervousness it would be a miracle if I got there alive.

The fear engendered by the mountains was palpable in the poems and writings of the Romantic artists who came here in the eighteenth and nineteenth centuries. The Shelleys, accompanied by Mary's stepsister, Claire Clairmont, arrived in Chamonix in July 1816 on an excursion from Geneva, where they were staying with Byron in a house he had rented on the shores of the lake. Like their fellow Romantics, the couple were terrified and inspired by the mountains. The snow-capped pyramids which shot up into the bright blue sky seemed to overhang their path, Shelley wrote in a letter to his friend Thomas Love Peacock. After seeing Mont Blanc for the first time he wrote: 'I never knew – I never imagined what mountains were before. The immensity of these aeriel summits excited, when they suddenly burst upon the sight, a sentiment of extatic wonder, not unallied to madness.' The valley, which they entered through a 'vast ravine', consisted of 'desolating snows... palaces of death and frost... avalanche torrents, rocks and thunder'.

The Shelleys climbed to Montenvers, on the eastern flank of Mont Blanc, where the jumbled crushed ice of the Mer de Glace tumbles over itself on its way down. Nineteen-year-old Mary Shelley wrote of its 'awful majesty' and described it as 'the most desolate place in the world'. Her experiences in Chamonix fed directly into the ghost story she was drafting, *Frankenstein*, which she would tell in the house on Lake Geneva, and which would be published the following year. 'I remained in a recess of rock, gazing on this wonderful and stupendous scene,' Dr Frankenstein, her narrator, says of an encounter with his monster in the Mer de Glace. 'The sea, or rather the vast river of ice, wound among its dependent mountains, whose aerial summits hung over its recesses... I suddenly beheld the figure of a man, at some distance, advancing towards me with superhuman speed. He bounded over the crevices in the ice, among which I had walked with caution.'

Another poet who was awestruck by the Alps was Wordsworth. He recounted in *The Prelude* how as a boy he was rowing across a lake when a mountain seemed to chase after him. The lake is Ullswater, in England, but it seemed to me to have been informed by larger and more sinister mountains.

> ... *a huge peak, black and huge,*
> *As if with voluntary power instinct,*
> *Upreared its head. I struck and struck again,*
> *And growing still in stature the grim shape*

Towered up between me and the stars, and still,

For so it seemed, with purpose of its own

And measured motion like a living thing,

Strode after me . . .

I had once found the Romantic poets' view of the Alps overblown, but now it had the ring of reportage. The rock-faces and *aiguilles* which filled this valley seemed hostile and powerful. The gravity of their enormous mass seemed to want to suck me in.

Our second night with Philippe was spent at 2,800 metres in a refuge built into a slope above the Argentière glacier. The next morning we were to climb 600 metres higher, over a pass that from the valley floor appeared to be a narrow, near-vertical slick of snow between two jagged cliffs, then across the glacier of the Trient plateau, over a second col, and down a long valley. As I thought about this col I found myself rolling over in my dormitory bed, shrugging off the blankets, eyes opening in panic. I needed to urinate. Groping for my head-torch and grimly climbing out of bed, I made my way out of the black room, down the stairs and through the unlit corridors to the three stinking, flushless, hole-in-the-floor toilets that served the needs of the eighty people in the hut. A warm wind, a föhn which had been blowing for the past several days, struck the building here, playing a mournful tune as it did so, accompanied by the percussion of a broken, slapping door. If the föhn continued in the morning, it would be hard to walk across the glacier, let alone climb, but somehow I knew Philippe would not call off the day's mountaineering. I was not going to find any comfort in this old wooden structure, and I spent a dreadful night walking between the stinking toilet block and the snoring masses upstairs.

Breakfast was at 5 a.m. and I watched enviously as *gesundheit* Swiss and Germans swilled coffee and munched cereal, their energy for the day. Away from the pistes and beyond the reach of the ski-lifts was a culture that was new to me. Here people spent days at a time touring above the snowline, staying in refuges like this with communal sleeping areas and no facilities except food, setting out before dawn across the glaciers and high mountain passes with ropes and harnesses and ice axes, to reach the next place before the snow thawed enough to avalanche. But where my introduction to skiing as a child had been a happy revelation, I was now fearful. I was surrounded by strong, energetic, unafraid people. 'Eat,' said Philippe. 'It is important.' I couldn't. I felt sick.

It was still dark when we left the refuge, cramponing down along a narrow strip of hard, cold snow by the light of our head-torches, towards the cache of skis and snowboards we had left the day before. As we reached the bottom of the descent, the massive shapes of the landscape began to reveal their jagged faces. Lights glittered halfway up the wall of mountains across the glacier. There were climbers up there on that impossible slope. They would have begun their ascents at midnight, working in the small patch of snow and ice illuminated by their head-lamps, creeping up towards their invisible goal. Madness.

Snowboards on, we set off across the glacier towards the approach to the col, Philippe setting a good pace at the front, me following with a heavy weight growing on my chest. The going was flat and hard for a kilometre, then the glacier steepened where it had thrown up great lumps of ice, and the route twisted and turned between and among the heavy blocks and crevasses. Each fall or stop was greeted with sharp words. '*Allez!*' 'We must be efficient now.' 'Stop, take off your board.' 'Walk.' 'Put on your board.' '*Allez!*' 'We must move quickly.' The white strip of the col and its steep-looking approach glared at me from afar, and the belief rose within me that this was a journey to self-destruction that must stop – but how? The granite giants seemed to be sucking me towards their precipices and rocky shoulders. I didn't want to look.

'I don't know if I can do it,' I said.

'*I* don't know if you can do it,' Philippe said icily.

My eyes were wet as I said, 'Stop!'

When we had skied back down to the bottom of the mountain, Philippe seemed to be in a strange state of ecstasy. It occurred to me that he might have been glad that I had refused at the first hurdle, that his beloved mountains had defeated the flatlander who had arrived with his snowboard and his plan to conquer them. He told me that people who worked in offices tried to complete famous routes like Chamonix–Zermatt, or climbing Mont Blanc, but they were not prepared to acquire the skills and techniques of mountaineering first. 'Mister Journalist,' Philippe said, 'ten metres at a time, the Chamonix–Zermatt is easy. But the globality of it, that is not easy.'

He would take us elsewhere, somewhere easier, to do some basic mountaineering, he announced. Nick and I had no Plan B. I was preoccupied with my failure: there was nothing now to complete, no journey to make, so why begin a new one at random? I wanted to return to Chamonix and sit in a café and try to work out what had gone wrong. But Philippe had his plan and we had the time and opportunity and we had paid a considerable sum, so we agreed to it. That afternoon we drove with him across the border into Switzerland, over the Col des Montets, past the turn-off towards the Grand St Bernard Pass which the ancients had used to cross the Alps, past vineyards and orchards that filled the rounded floor of the upper Rhône valley.

I phoned Lucy from the car. I hadn't spoken to her for several days, and when she answered I felt a lump in my throat.

They were in the garden. It was a hot spring day. I could hear the children in the background, playing and laughing with their grandmother. They had been to the zoo, Lucy said, and now they were painting.

'I've pulled out,' I said.

The myth that I had constructed and chosen to believe, of myself as the winter sportsman who couldn't live without his fix of snow and physical challenges, lay in pieces. I wanted to be at home, in the garden, with them.

The story of the modern ski movement can be traced through the interlinked lives of three men: a famous Arctic explorer, a bloody-minded Czech-born hermit and a poor Norwegian peasant.

It begins with the peasant, Sondre Norheim, who was born into a poor family in the valley of Morgedal, in the mountains of Telemark in southern

Norway, in 1825. Telemark in the nineteenth century was a region of rural communities and tenant farms, many of them occupying unfertile or steep land that was unsuitable for ploughing. Getting around the remote valleys was difficult, particularly in the snowbound winter months. The horses and carts of the wealthier farmers were beyond the means of most. So the people of Telemark developed a great facility with skis. Skis were available to everyone; all that was needed was the skill to cut the wood and sculpt it into shape. In winter they would be used for fetching and carrying on the farm, as well as for pleasure. Children would be put on skis from the age of three or four, and people would ski right into old age. Competition was intense, and on Sundays the young men would gather to demonstrate their skills on the fastest and most difficult runs, while the young women, who also skied, egged them on. Jumping from roofs was a particularly popular test.

Among the large number of skilled skiers who grew up in Telemark, Sondre Norheim would become by far the greatest. In his teens, he discovered the principles of what we now know as Olympic ski jumping. He realized that landing on a slope, rather than on the flat, would enable him to cover much greater distances. It looked more impressive too. He was a daredevil and a showman, taking on the most difficult runs and betting with his peers whether certain stunts or jumps were possible. On one occasion Norheim skied at great speed through a sawmill, ducking the whirling blade at the very last moment. It helped that he made his own skis and bindings. It has been said that he was the first person to attach the heel to the ski as well as the toe, which gave him much more control over turning, and to sculpt a waist in his skis so that the tips were wider than the middle. Whether these innovations were really his to claim we may never know, but by using them so well and so famously, he introduced them to the skiing world.

During the 1860s, Norheim's reputation rapidly spread beyond his own small valley and in 1866 he was invited to take part in one of the first ski-jumping competitions at Ofte, Høydalsmo, fifteen kilometres west of Morgedal, where he won first prize. Two years later, now at the ripe age of forty-two, he had his biggest triumph, in the Norwegian capital of Christiania. The Christiania competitions were among the first civilian games ever held for the sport, and Norheim, even after skiing the 200 kilometres from Morgedal, took the top prize with ease. More than that, by showing what could be done on skis, he gave the sport a tremendous fillip. The *Aftenbladet* newspaper report on 10 February 1868 read: 'There was something so remarkable about [Norheim's] racing and his movement on skis that one had to believe that for him it was the natural way to move.' Norheim was full of faith in his own ability, to the point of arrogance. Asked what he made of the Christiania courses, he said he 'could have skied them with a sack of potatoes on my back'. The skiers of the capital were so impressed that they gave names to two of the new turns Norheim had demonstrated: one would ever after be called the Telemark turn; the other the Christiania, later shortened to Christie. His achievement is all the greater considering that he was twenty years older than many of the other skiers. He entered the competition again in 1871, when he came second, and in 1875, when he was forty-nine, he still came a respectable sixteenth.

The first superstar of skiing made no fortune from his celebrity. Although he became a folk hero, he remained poor for most of his life. In 1884 he emigrated to the Dakotas, where he died in 1897. The flame of the Winter Olympics has three times been lit in the Morgedal cottage where Norheim was born, in honour of the debt that skiing owes him.

The national celebrity of Norheim and the Telemark skiers who followed him led to an explosion of skiing in Norway, particularly in Christiania. Among Norheim's many fans was the second of our ski pioneers, the explorer Fridtjof Nansen. If Norheim had made skiing fashionable in his home country, it was Nansen who would take it to the world.

Nansen was born in 1861 into a prosperous family on a farm at Store Frøen, a rural area north of Christiania. His mother encouraged her children to take strenuous physical exercise, including skating, swimming and skiing. Nansen, who was four when he tried on his first pair of skis, later described the electrifying effect of the Telemark skiers on Christiania, writing in 1890 that where ten or fifteen years earlier the hills and forest paths around the capital in winter had been empty and desolate, now on a fine winter Sunday they were thronged with skiers of all ages and physical ability. Nansen wrote: 'A well-known *skilober* [skier] from Telemarken, Söndre Auersen Nordheim by name, is reported to have jumped... ninety-nine English feet from a projecting rock and to have kept his balance when he alighted below.'

In 1883, news reached Nansen of the exploits of two Sámi skiers who had been exploring the uncharted interior of Greenland. Several expeditions had tried and failed to cross the Greenland ice cap, including a British one led by the conqueror of the Matterhorn, Edward Whymper, and the myth persisted that there was an oasis of fresh water and plants to be found among the harsh

glaciers and mountains. One autumn evening, Nansen was sitting listening

indifferently to the day's paper being read aloud, when, 'suddenly', he recalled,

'my attention was roused by a telegram which told us that [the Swedish

explorer] Nordenskiöld had come back safe from his expedition to the interior

of Greenland, that he had found no oasis, but only endless snowfields, on which

his Lapps were said to have covered, on their "ski", an extraordinarily long

distance in an astonishingly short time. The idea flashed upon me at once of an

expedition crossing [Greenland] on "ski" from coast to coast.'

The exploits of Nordenskiöld's Sámi skiers were indeed extraordinary.

In one fifty-seven-hour period, Pavva Lásse Tuorda and Anders Rassa had

covered 460 km of the Greenland interior on skis. The disbelief that met

Nordenskiöld's claim was such that he organized a 220-km race in 1884 to prove

that such exceptional distances were possible. Pavva Lásse Tuorda, then thirty-

seven, won it in 21 hours and 22 minutes. Five of the first six finishers were

Sámi. The point was proved.

Having seized on the idea of taking skis, Nansen began to make plans for

his own trip to the unknown interior of Greenland. Other expeditions had

started on the inhabited west side, but Nansen decided to begin his journey

from the east coast. The Norwegian public was sceptical and thought he was

reckless. Nevertheless, in May 1888, Nansen and his team, including the Sámi,

left for Greenland by way of Scotland and Iceland. On 15 August they began

their crossing, and a little over forty days later, on 26 September, arrived

on the west coast. Nansen returned to Christiania the following year a hero.

Although he was still a young man, he had already made a reputation as one of the greatest Arctic explorers of all time.

Nansen lost no time in spreading the word about skiing, for which he claimed 'a position in the very first rank of the sports of the world'. The book he wrote about his expedition, *Pá Ski over Grønland*, became an extraordinary hit in Norway and was translated into many European languages, including English in 1890 and German in 1891. It included a paean to the skis that were such an important piece of equipment, but which were so unfamiliar to the non-Norwegian audience that a footnote had to be supplied telling readers that a ski was 'literally, a "billet or thin slip of wood"... the Norwegian name for the form of snowshoe in general use among the northern nations of the Old World'.

Nansen knew, he wrote, of 'no form of sport which so evenly develops the muscles, which renders the body so strong and elastic, which teaches so well the qualities of dexterity and resources, which in an equal degree calls for decision and resolution, and which gives the same vigour and exhilaration to mind and body alike. Where can one find a healthier and purer delight than when on a brilliant winter day one binds one's "ski" to one's feet and takes one's way out into the forest? Can there be anything more beautiful than the northern winter landscape, when the snow lies foot-deep, spread as a soft white mantle over field and wood and hill?'

For Nansen, skiing was part of the Norwegian psyche. He wrote of the Norwegian peasant childhood how snow lay soft and deep outside the cottage door the whole long winter through, how children would practise constantly from the age of three or four, how, on Sunday afternoons, all the youth of the parish would meet on the hillside to outdo one another at skiing, as long as the brief daylight lasted. Nansen would probably have been as surprised as everyone else when the next great skier turned out not to be a Norwegian, or even from Scandinavia at all.

Matthias Zdarsky read a translation of Nansen's book in his remote mountain retreat at Hebernreith near Lilienfeld, some seventy kilometres south-west of Vienna. In 1889, at the age of thirty-three, Zdarsky had moved away from the hustle and noise of Austrian society in order to develop his 'scientific–artistic ideas' undisturbed. In photographs he appears small, wiry and athletic (he had been a gymnast), with either moustache or beard or both. By all accounts, Zdarsky was a determined, belligerent man, quick to see the faults of others and point them out, who in time would create for himself the role of prophet of the new sport and make some powerful enemies as a result.

Nansen's account of life on the Greenland ice cap launched Zdarsky into a frenzy of skiing. He ordered a pair of skis from Norway, and travelled back on them from the post office where he'd picked them up. They were enormous, in the Norwegian style, nine foot six inches long, with toe bindings that gave no lateral stability at all, and though he made his way up the gentle slope out of town on them with some facility, the steeper slopes gave him trouble and turning was almost impossible. But even as Zdarsky climbed his way back towards Hebernreith he was working out how to improve his new acquisition, to adapt them to the steep mountains around Lilienfeld.

Skiing was still in a very primitive form in the Alps. Arnold Lunn relates how his father, also a travel agent, brought a Swiss ski teacher to Chamonix to

teach his clients in the winter of 1888/9. 'Somebody asked him if it was possible

to turn. He replied in the negative, but added that a long gradual turn was just

possible if one dragged oneself around on the pole. He claimed to have seen an

expert perform this difficult manoeuvre, but modestly added that he was unable

to demonstrate it himself.' Arnold's son, Peter, recalls his father spending three

winters skiing in the Alps before he saw anybody do a turn. Instead skiers sat

astride their single poles like witches on broomsticks, so that their weight on the

pole exerted a braking effect. Other written manuals were not much help. The

journal of the Austrian Ski Association gave the following advice: 'On the

descent the ski-runner leans back on his stick and shuts his eyes. Then he darts

downward straight as an arrow, and continues till he can no longer breathe. He

then throws himself sideways on the snow, and waits until he regains his breath,

and then once again hurls himself downwards till he once more loses his breath

and throws himself on the snow.'

Alone and isolated from the growing skiing community, Zdarsky was

working out how a sport that had been incubated in the low and rolling hills of

Scandinavia could be adapted to the precipitous Alps. He skied in the mountains

around Lilienfeld for the next six winters without once seeing another skier,

and made numerous adaptations to his skis, trying at least two hundred different

types of binding, and cutting them down from 294cm (9ft 7in.) to a more

manageable 220cm (7ft 2in. – still long by today's standards). He discovered

that by putting his weight on the uphill ski while keeping the downhill ski flat,

he could accelerate it around the outside of the turn and end up traversing the other way across the slope. In other words, he worked out the principles of the stem turn, and by linking them in quick succession, was able to slow his descent on even steep slopes. By the end of his six-year exile, he believed he could descend almost all of the terrain around Lilienfeld.

In 1896, the man known to Arnold Lunn as the Hermit of Lilienfeld broke his retreat. He set off to watch a ski competition in Semmering to see how he measured up to other Austrian skiers, and was astonished to see them braking by leaning back on their single poles on what we would now call the nursery slopes, while the whole mountainside was left untracked. When he told the leader of the local ski group that it was possible to ski on the steep slopes too, the man tapped his head with his finger, indicating that Zdarsky was mad. He had not brought his skis with him, so could not prove his point. Later that year, however, he published his book, *Die alpine Lilienfelder Skifahrtechnik*, the first ski instruction book. It became one of the best-selling ski books of all time, running to a seventeenth edition.

However, Zdarsky's arrogance and lack of social grace alienated him from many in the Austrian skiing community. Who was this man who had never competed in any races, had never crossed the ice caps of Greenland, had never been to Norway, the home of skiing, or even studied the Norwegian style? And how dare he come along with his homemade bindings, short skis and stem turns, and tell them what to do? A war of words broke out between different skiing factions, and Zdarsky was not slow in helping it turn personal. After a Colonel Georg Bilgeri made the unfortunate remark that one used a 'hind leg' while skiing, Zdarsky tartly commented that there appeared to be an officer of the

Imperial Army who had four legs. He was promptly challenged to a duel by the colonel.

As Zdarsky's fame spread, pupils travelled to Hebernreith to learn from him. He gave his expertise freely, and his didactic style was suited to telling people what to do. 'There must always be a listener and a talker' was his view of teaching, and Zdarsky was usually the talker. One of Zdarsky's disciples, W. R. Rickmers, wrote of visiting him for lunch and being given a lecture on 'the noble art of making tea'. 'The theory and practice of the production and consumption of this beverage is one of Zdarsky's hobbies, but as such only an extension of his very serious views on temperance, or rather abstention for mankind in general and the ski-runner in particular... Such was Zdarsky: campaigner and preacher, inventor, theorist and artisan, rolled into one.'

His control of his pupils was absolute – to the extent that he would teach groups of up to two hundred people, and once, when he was drafted into the Austrian Army as a ski instructor during the First World War, a group of 1,600. On Sundays, special ski trains were run from Vienna to Lilienfeld for people who wanted to learn under his instruction. His expertise was respected even if his disciplinarian techniques were frowned upon by clubbable British skiers. Lunn described Zdarsky's acolytes as 'a motley army of students, clerks, girls and matrons, elderly civil servants and doctors who poured out of Vienna every Sunday to receive the gratuitous instruction which Zdarsky so freely and so generously offered to all who would submit to be drilled like a squad of soldiers'.

Predictably, the Norwegians themselves took a lofty view of Zdarsky's teachings. Eventually, in 1905, the Norwegian National Association sent one of its best skiers, Hassa Horn, to Lilienfeld to assess Zdarsky's techniques. Herr Horn reported even-handedly that, although he judged himself to be swifter on the flat and across country, Zdarsky was the most expert skier he had ever seen on very steep terrain. He added, prophetically, that the Norwegians would have to watch Zdarsky's Alpine techniques, lest Norwegian skiing be overtaken.

Zdarsky spent much of the latter part of his life in pain. In February 1916, while he was instructing the Austrian Army, an avalanche that few others would have lived through overtook him. A friend of his noted that 'He was literally smashed, crunched, kneaded and compressed by the snow, and almost every part of his body subjected to enormous strain. Yet he survived, but as a medical curiosity. He had about eighty fractures and dislocations, among which were half a dozen dislocations of the spine. With characteristic energy and patience he set about curing himself and inventing apparatus for his limbs, suffering agonies all the while. He is quite crooked and his legs twisted. But he wrote to me that he has managed to ski again.'

By the time of Zdarsky's death in 1940, skiing had taken root in the Alps. In 1924, the first Winter Olympics had been held in Chamonix, and after the Second World War a method of travel that had begun in the Stone Age among the snowbound tribes of central Asia would become a modern, global sport.

His memorial stone in Lilienfeld carries a quotation from the once sceptical Arnold Lunn: 'Zdarsky will never be dethroned from his position as the father of Alpine skiing.' Wrote Lunn: 'Zdarsky was the Moses of the new religion.'

✳

East of Sion, Philippe turned off the main road and drove upwards through a narrow gorge into the town of Leukerbad, set beneath a stunning, 600-metre-high, crescent-shaped cliff-face which blocked all points north of the small town. Over lunch, Philippe showed us a cutting from the previous Saturday's *Le Dauphine*, a profile of himself, a great mountain man, telling of his feats in the Alps and in Nepal. In my misery, I wondered if he thought he was wasting his time with us.

The cable car shipped us to the top of the cliff in the early afternoon. Up here above the world was a small, high valley with a frozen lake at the bottom and higher peaks all around. The sky was clear blue and we looked across at the great mountains to the south, to the route we had intended to take, through some of the most famous peaks in the world: Mont Blanc, Monte Rosa, Weisshorn, the Matterhorn. Philippe talked us through them from this remote distance, as if he knew every crevice.

Here, in the high valley of the Wildstrubel, mountaineering boot camp began. The days were characterized by a 6 a.m. departure and a fast ascent, with no time to linger. 'Back there, you asked to stop,' Philippe would say, halting after an interminable climb. 'Why?' There was never a good answer to this question, as nothing was more important than ascending while the snow was still firm. The panoramas were neglected, and instead I learnt the mountaineer's trick of shrinking the world into the tight little corner around my feet.

Philippe would select a line in the snow which would cross the slope but also move up it, if possible along the tracks made by an earlier skier – the snow was always firmer there. Along this line, the guide's crampons dug small prints into which I, in the middle, would try to plant my own cramponed boots, and to the side my ski poles, to get a grip in the snow. Each step was carefully placed, as there was almost always a steep drop close by, and even when there wasn't, my imagination was conjuring a precipice. We attached our snowboards to a cord, which in turn was attached to our harnesses, and dragged them behind us, even on the steepest slopes and traverses, because it was easier than carrying them, and it kept them out of the wind. We were often tied together by Philippe's thick climbing rope, which tautened and slackened according to the varying speeds of the three people in the chain. Now and then, he would switch back on himself as he plotted our climb, and the rope and the cords would need to be stepped over, untangled or adjusted so that they hung on the other side. We would slow down for this, but not stop. At other times, when the going was flat, we would strap on our boards and use our ski poles to punt us along, depending on the condition of the snow.

As the sun grew warmer and our packs became heavier, sweat ran off our faces and a powerful body smell emanated from our unwashed clothes. On the steepest climbs, sun cream slid down our dripping faces through our eyebrows and into our eyes. All the time the slow plod of step after step, following Philippe's steady pace: pole, boot, pole, boot, pole, boot, pole, boot… We continued in this way for hour after hour, as fragments of the same repetitive pop tunes ran around in our heads.

Each day was a slow race against the temperature. The snow was hard in the mornings and terrible to ski on but good for crampons, while from 2 p.m.

on, it was too soft to walk through with ease, and our feet would slip on the steep ascents. There were avalanches from mid-afternoon into the early evenings. But no matter what time of day, the snow was very different depending on where it lay on the slope, and how exposed it was to the sun. Philippe would pick a route up the hardest snow, in the shadow of a peak, or where the sun struck it obliquely. When we had to walk near snow that was exposed to the sun we would hurry past, Philippe cursing us for being slow, for dropping our gloves or wanting to take photographs. By mid-afternoon, as we returned to the refuge, snow that had been hard ice in the morning had turned into a peat bog, a thick soup into which even a snowboard would sometimes sink several feet. We would drop our gear in a heap and sit outside the hut and watch the snow. Sometimes we could see it avalanche off the mountain, tumbling with a roar and a hiss near one of the routes we had taken uphill.

The mountains were different every day: the land was more exposed; the open, thawed part of the river was growing; more rock was showing. When we stopped on top of a col or on a peak and looked down into the valley, the line where green met white had moved, almost imperceptibly, a little higher than it had been the day before.

Although the setting was idyllic, the sense of alarm I had felt on the Argentière glacier had not left me, and the mutual suspicion that had been sown between Philippe and me on our first day had hardened into dislike. One day, as we approached the summit of the Daubenhorn, he turned to me. 'You fear death, Mister Journalist,' he said. 'Why?' We were resting below a ridge at the highest snow-covered point on the mountain, and the path he had chosen for us was rocky and precipitous. I had told him I did not wish to follow it, and would stay there until he and Nick returned. He interpreted my obstinacy as a great failure on both our parts, but mostly on mine. In an attempt to lighten the mood I said that I wanted a few more years yet, thinking that I had some things left to do – bringing up the children, repaying the mortgage. But mainly, I told him truthfully, looking up at the path composed of loose shale that he wanted me to take, I was scared. My hands and legs shook. When I thought about the murderous drop that lay just behind the ridge, I could feel an electric sensation running down the backs of my wrists towards my fingertips.

'I know you are frightened of the mountains,' Philippe said. 'I understand that. Maybe, if you make a mistake and fall, you die, yes. But didn't you come here to challenge yourself?' I made a face. 'I think you are dead while you are alive,' he said. 'More and more, I think, society is made up of people like you. You take risk unconsciously. When you are in the town, or driving your car, you take risk but you don't think about it. Now you are with me, and this is a conscious risk, you say you will not take it. But if you do not come, you will feel bad. Will you take it?'

I shook my head and stayed put.

Nick walked off with Philippe along a path that narrowed to barely a metre wide, and which fell away sharply on either side, to the cross at the summit. I waited by the rocky trail.

Soon after that we drove back to Chamonix. In the car park we took from our rucksacks the crampons, shovels, probes and transceivers that Philippe had loaned us, handed them back to him, and shook hands briefly. I lied when I said

I had enjoyed it, but I needn't have bothered. He wasn't interested. He grunted a half-hearted 'Welcome to the mountains,' and was gone.

That night Nick and I went out on the town and got drunk. We talked about Philippe and the things he had said, and tried to work out where exactly it had all gone wrong.

For a long time afterwards I saw myself through Philippe's eyes. I realized that the fear of falling, of taking a risk and losing everything, had been slowly creeping up on me. Beyond the odd foray into snow, the adventures I had sought as a younger man no longer tempted me. Instead I had become cautious. I warned the children to stay back from the kerb, I worried about how much television they watched, and what they ate. I had become complicit in our risk-averse society. Philippe, meanwhile, who had lost many friends in the mountains, still returned to them each day.

Everyone must draw their own line which they will not cross. I had drawn mine, and it lay far short of where I had expected it to be. I felt hollow.

I thought about my father, and how I was nearing the age at which he had died. To try to escape a particular bout of depression, he would change something in his life. My parents moved right across northern Britain, as he took different jobs, from Newcastle to Edinburgh, Hull and York. It must have been harder to move when my brother and I were born, so he changed the house instead. He dug a swimming pool in the back garden, laid a huge patio, built an extension and a summerhouse. He built walls, and even a ha-ha. When he'd done that, he changed the car. In the end, there was nothing left to change except his family.

Sometimes we follow the same paths as our parents, and can't help it. We fall into the same traps.

When I was in my twenties and fled London for the moors, I would follow the long-distance footpaths set out in the books of Alfred Wainwright. On a particularly bleak mountainside where the rain came down in rods and people's thoughts turned to the comforts of home I remembered he exhorted the reader to continue: 'Don't give up. If you give up now, you will be giving up for the rest of your life.' Back then, I had not given up. Wainwright's words now seemed to mock me: unless I made a conscious effort to break it, giving up could too easily become a habit.

In the Alps, all that separates the climber from the bottom of the cliff is a short slip sideways. Whenever I shut my eyes I saw myself on the knife-edge ridge near the summit. I was trying to concentrate on the way ahead, but the void kept beckoning me.

'The mountains find your weaknesses,' Philippe had told me. '– the ones in your mind. You must work on them. Then you can come back.'

CHAPTER 6

An Evil Element

Lewes

I caught some indeterminate fever on the journey back from Chamonix. Shortly after I arrived home I took to my bed and stayed there for a week, unable to get up or sleep properly. The ceiling of our bedroom is decorated with Victorian cornices which have become clogged with successive layers of white paint. These meringue projections now took me back to the Alps and the wind-blown snow and ice formations and glistening peaks that had preyed on my mind. I was walking again in the Wildstrubel, where snow fizzed from the cliffs around the Lämmerenhütte and landed a hundred metres below with a crump like a mortar round. I envisaged myself caught in one of these waterfalls, knocked over and smashed, blind and tumbling like poor Matthias Zdarsky, who could feel the power of the snow breaking his bones, squeezing him until he thought his eyeballs were bursting from his head, forcing his ribs to grate against his spine. The Sublime was now an unbalanced concept for me: it was too close to terror.

That week summer arrived in London. The rangers filled the giant concrete paddling pool in our park and the sound of shouting and splashing children drifted in through an open window on the warm wind. As Lucy and the boys went out into the sun I struggled through the drifts in my mind.

I was angry with Philippe, but more with myself. With his help I might have wrestled down my fear and we would have made it to Zermatt. Clearly, he was not one to encourage his clients, but why had that mattered? Was I such a slave to confidence? Why had I not argued with him at the top of the Daubenhorn, or left him and found someone else to guide us? I chose instead to withdraw, to relinquish interest and control of the journey in which we had invested so much. Before we arrived in Chamonix I had believed that finishing the *haute route* would open up a glorious future of ski mountaineering across the roof of the world. Instead the opposite had happened. I had reached a terminus, a full stop.

I thought about the young man I had met in the Alps two years ago who had been killed by an avalanche the following day. He was a ski rep and an experienced snowboarder who had been lured off-piste on one of those sunlit mornings when six inches of fresh snow make your heart rise up in your chest. His friends had walked around ashen-faced for days after they heard the news, and when they spoke it was only to say what a waste it was. It had been a freak event, a lightning strike, someone had said; but snow is much more likely to kill than lightning.

I looked back at my own skiing history and attempted to calculate how close I had come to disaster, but it was impossible to tell, even in retrospect. I remembered a day I was caught in a blizzard in Switzerland. It had been snowing steadily for twelve hours and the fresh flakes were blown into deep drifts on top of a hard, icy layer. After we had left the piste the wind had strengthened, whipping the fresh powder into an opaque soup. We could no longer see more than a few feet. I sat on the downhill lip of a flattish mountain road and looked back towards the rest of the party, but their familiar colours and bulky shapes had disappeared into the cloud. They could not have been far away, but by the time I climbed back up and across to where I thought they should be, they would have moved on. Anyway, I wanted to be down the hill first, so I set off alone.

The wind blew snow off the windward side of the ridge and carried it into the broad gully we had been travelling down. I could no longer make out the detail of the ground in front of me, or see the rocks that would have given me a sense of where I was. I fell badly almost immediately, and lay winded, staring up into the swirling white as I caught my breath and worked out if I had hurt myself. I could hear nothing above the battering of the wind on my hood and my breath, sucking and blowing through the faceguard of my jacket, making my clothes, which were already damp from sweat, wetter than ever. When I lifted my head I could see the flakes drifting up against my dark shape, gradually painting me out. If I lay there long enough, I would be buried completely.

Before my T-shirt and jacket lining had time to freeze I struggled to my feet and attempted to descend again. There was no meaningful sound, I could see little, my sense of touch was blurred by the softness of everything. I couldn't even tell which way was downhill. I followed the memory of the direction I thought I had been travelling in, tumbling over every few feet, straining to push myself up as my hands sank into the snow. I floundered along for a long time before I saw the orange stripe on top of a piste marker, and salvation. I made my over way to it, and from there I could make out the next one, and then the one after.

When I arrived at the bottom of the mountain the others were waiting. They had descended easily by another route, and were angry that I hadn't waited and stuck with the group. They had almost called the mountain rescue. Don't be so soft, I had said. It seemed now to have been another era, when I was a different person.

In Japan, blizzards were once believed to be manifestations of the snow spirit, Yuki-onna. She appeared as a young woman with pale skin and a kimono of whitest white. She often had a baby in her arms, which she asked travellers to take from her, but if they did they would find that it was actually a lump of hard ice, which would freeze them to death. A traditional Yuki-onna story tells of two woodcutters, Mosaku and Minokichi, who took shelter from a snow-storm in a ferryman's hut one night when they were unable to cross a river to reach home. Old Mosaku fell asleep almost immediately, but Minokichi lay awake listening to the snow battering at the door as the river roared and the hut swayed like a junk at sea. At last, Minokichi also fell asleep, but he was awak-ened by a showering of snow on his face. He looked up to see a woman bent low over Mosaku, blowing her breath over him, which appeared like a bright white smoke. Then she came towards Minokichi until she was almost touching him. He saw how beautiful she was, though her eyes scared him, and he found he was unable to move or speak. After a while she smiled at him and whispered, 'I intended to treat you like the other man. But I cannot help feeling some pity for you because you are so young. You are a pretty boy, Minokichi, and I will not hurt you now. But if you ever tell anybody what you have seen this night, I will kill you.' By dawn the storm was over, and when the ferryman returned he found Minokichi lying senseless beside the frozen body of Mosaku.

A year later, Minokichi fell in love with a woman and married her. Over the years, the couple had ten children. Then, one night, Minokichi told his wife about the night he had spent in the ferryman's hut, and the woman he had seen, and how she resembled her. Suddenly, his wife dropped her disguise and revealed herself to be Yuki-onna. She flew into a rage, shouting, 'But for those children asleep there I would kill you this moment! Now you had better take very, very good care of them, for if they ever have reason to complain of you I will treat you as you deserve.' And she melted into a bright white mist and was never seen again.

Greek myth is similarly suspicious of snow. Chione, the snow-nymph, is another *femme fatale*. She makes love to Poseidon and has a son, Eumolpos, but is so terrified of the reaction of her father, Boreas, the god of the north wind, that she throws the baby into the sea to avoid discovery and he survives only thanks to the quick thinking of Poseidon.

In the Alps, the aspect of snow that generated the richest source of myth and superstition is the avalanche. The white waves that could sweep unheralded through towns and villages were thought to be created by dragons, witchcraft, malevolent spirits or an angry God. If one house was left standing while others were demolished, it was not because snow is indifferent to where it lands, but because the evil guiding spirits favoured the occupants of the surviving house, who by inference were thus up to no good. These spirits were sometimes said to have been seen steering avalanches with a tree, like a rudder. Clairvoyants, magicians and soothsayers would be summoned to get rid of them.

Until the eighteenth century, most of the inhabitants of the Alps avoided going above the snowline at all for fear of the demons and dragons who lived there. Mont Blanc itself was known until the mid-1700s as Mont Maudit, 'the accursed mountain'. Those who crossed the high Alpine passes generally did so because they had no choice: they were transporting goods for trade, going to hunt or herd the animals, or making pilgrimage. Many were blindfolded to prevent them from being overwhelmed by the scenery. In the 1720s, when Johann Jakob Scheuchzer, a professor from the Carolinum in Zurich, wandered the Alps collecting theories about the nature of the mountains, many people were prepared to tell him of their face-to-face encounters with dragons and fairies. Dragons came in a variety of shapes and sizes, Scheuchzer reported, and had a jewel in their foreheads which, if it was cut out while the dragon slept, could cure a variety of ailments.

One legend from Erstfield, in the Swiss canton of Uri, told of an old woman dressed in black who was seen riding the first wave of an avalanche while quietly turning her spinning wheel. The woman was grabbed by four men and burned alive. There was also the famous trial at Avers in the Hinterrhein valley in 1652, at which it was baldly stated that 'witches are the cause of avalanches'. Inhabitants of the canton of the Grisons would protect themselves from such devilry by burying eggs marked with the sign of the cross at the foot of known avalanche slopes.

The avalanche historian Colin Fraser recounts an Alpine adage that sums up the mountain-dwellers' fear of snow: 'What flies without wings, strikes without hand and sees without eyes? The avalanche beast!'

After the fever passed, I tried briefly to return to my books, but I was listless and couldn't concentrate. The desk where I had spent so much time was now covered with household flotsam: clothes, newspapers, and duvets that hadn't quite made it into the airing cupboard. The notes and files, never tidily kept, now lay disorganized and gathering dust: the snow project had lost momentum. Instead there were other things to occupy me. I was busy at my day job, and later that summer Eddie was born, and we were plunged again into the business of caring for a new baby. It wasn't until mid-September that I thought about snow again, and that was because of a strange coincidence.

One of our first ventures beyond London with our new son took us to Lewes, the county seat of East Sussex. Lucy had shown a passing affection for this pretty hilltop town and I was trying to fan this spark of interest into an enthusiasm that would lead to us moving there. Perhaps because of its geography – it lies at the conjunction of rolling downs, a narrow river valley and several ley lines, and is close to the sea – Lewes is an eccentric place, with a history of paganism and radicalism. It is home, for example, to the head-quarters of the world's largest druid order, and in 2005 the people of Lewes elected a witch to the town council, a Liberal Democrat one at that, and later promoted her to chairman. On the Fifth of November every year Lewes's bonfire societies compete to burn an effigy of Guy Fawkes or Pope Paul V. The town also has a history of strange weather incidents.

We walked from the Norman castle down the hill past the house where the eighteenth-century pamphleteer Thomas Paine had run a tobacco shop before getting caught up in the American Revolution. We walked into and out of Woolworths, and across the bridge over the River Ouse. At the lower end of South Street a dual carriageway and a chalk cliff blocked our route, so we turned back towards a nearby pub, the Snowdrop Inn, for lunch. It was one of the hottest days of the year, and in the beer garden the tables were stacked with empty glasses, the ashtrays were full and the sun was scorching down on the palm tree that stood in the centre of the hot gravel. After the brightness outside, the interior of the pub seemed almost black. As my eyes adjusted to the gloom I found my attention drawn to a framed, yellowing newspaper cutting on the wall. 'The Lewes Avalanche, From the Memoir of William Thomson Esq.,' it read, and went on to tell the story of how the Snowdrop Inn got its name.

Thomson begins by describing the weather in late December, 1836. The snow had begun to fall on Christmas Eve, when a storm hit South-East England, and didn't stop until Boxing Day. The quantities that landed had been seen before in this corner of England, but this time gale-force winds drove the snow into drifts up to twenty feet deep in places, and the streets were covered in a white layer five to ten feet thick. The roads from Lewes to London and Brighton were blocked for almost a week, even though gangs of workmen were engaged to dig a route through for the mail coaches. One Lewes resident recorded on Christmas Day that 'everything is most dreary and wretched – I returned to my den this evening – wet through from walking in snow up to my knees'. The snow blocked his chimney, he wrote, and the smoke came down instead of going up. Thomson had struggled home from Great Yarmouth via London through deep drifts, arriving in Lewes just before midnight on Christmas Eve. 'On reaching my residence in the centre of the town, I found the snow had drifted over the front door,' he wrote. 'On its being opened, it fell inwards and froze so hard and rapidly to the door post that for nearly an hour the servants were unable to close the door; sweeping had no effect, the icy particles were obliged to be scraped from the woodwork.'

In 1836, the eastern side of South Street where the Snowdrop Inn now stands was occupied by eighteen cottages, built for the families of poor workers and known as Boulder Row. They backed on to a chalk cliff-face, 100 metres high, which had been created by quarrying into the Downs. On Christmas night, a violent gale blew the heavy snowfall into a cornice at the cliff's edge, ten to fifteen feet thick, above the cottages below. 'Tons upon tons seemed to hang in a delicately turned wreath as lightsome as a feather,' the *Sussex Weekly Advertiser* reported later, 'but which, in fact, bowed down by its own weight, threatened destruction to everything beneath.'

The inhabitants of Boulder Row appear to have been transfixed by the danger above. Even on the night of 26 December, when a portion of the snow crust fell from the cliff-top into Charles Wille's nearby timber yard, lifting his sawing shed and depositing it in pieces forty feet to the west, they refused to leave their homes. Early the next day, cracks could be seen in the cornice and several people again asked the cottages' residents to move. Some now did, but more remained, while some of those who had left kept returning to collect furniture, clothes and food. At 10.15 a.m., Robert Hyam of The Schooner beer shop opposite Boulder Row grew so alarmed that he tried to drag two women bodily from their homes, but eventually left without them, fearing for his own life.

The *Advertiser* talked of 'some fatal infatuation' with the snow among the remaining residents. It was either that or simple bloody-mindedness: 82-year-old William Geer said that he had lived safely beneath the cliff for many years and would not budge now. He was later found halfway across the road, the timber beam that killed him lying across his body. Hyam, meanwhile, had just got clear of the cottages when a stretch of the cornice gave way and fell on the last seven houses in South Street. One eyewitness said the snow appeared to strike the houses at the base, heaving them upwards, then breaking over them like a gigantic wave to dash them bodily into the road; and when the mist of snow, which then enveloped the spot, cleared off, not a vestige of habitation was to be seen – there was nothing but an enormous mound of pure white.

Having set out for a walk on the Downs on the morning of the 27th,

Thomson was stopped by the agitated solicitor John Hoper, who told him that
the snow had fallen and that many were dead. Thomson hurried to the spot:
'From what I could learn (for there was little to be seen but snow), I found that
seven cottages and their inmates were buried,' he wrote. 'The force of the
descending mass of snow had absolutely driven the cottages from their foundation and carried them nearly across the public road, about 35 feet wide at this
part.' Evidently the right hero for this hour, he ordered dozens of shovels from
a nearby ironmonger and got the neighbours digging for survivors. He could
see that the cornice had not fully dropped, and told other men to cut through
it, while he kept a look-out to warn the workers below, should any further
avalanche fall. It duly did.

This moment is portrayed in Thomas Henwood's painting *The Lewes
Avalanche*, one of several items of avalanche memorabilia kept in a house once
owned by Anne of Cleves and which now serves as a museum. It isn't a great
work of art, but it is full of narrative. In the foreground, a horse rears with
bulging terrified eyes as the second deluge cascades from above. The beams of
the demolished cottages fill the road. In the centre of the image, the body of old
William Geer lies crushed under an oak beam. A group of ghoulish onlookers
are held back several hundred yards up the street. Thomson himself, in a top
hat, stands at the left of the painting and gestures to the cornice. On his shouted
warning, workmen who have been digging with pickaxes among the flattened
houses, which have been reduced to spars, crumbled chimney breasts and
segments of roof, begin to flee the new snowfall. The sky above the ridge is a
furious black, and in the centre of the image four men carry a woman on a

stretcher, their arms around each other's shoulders, like pall-bearers. Eight people died in the first avalanche. Thomson was buried to his shoulders in the second, but no one was killed by it.

The story on the wall of the Snowdrop Inn ends with Thomson recording that seven of the dead were interred in a single grave in South Malling church-yard on the following Saturday. They had to cut a route through the blocked streets so that the wagons could pass, so deep was the snow. A marble plaque in the church remembers the day 'the Poorhouses of this Parish was [*sic*] destroyed by a Mass of Snow Falling from the Hill Above'.

The Lewes 'snowdrop' remains the most destructive avalanche ever to have fallen in Britain. As I carried the tray of drinks into the sweltering garden, I looked up at the cliff behind the pub. In midsummer it was hard to imagine the frozen white death plunging down on this county town.

Back home in London that evening, I cleared a space on my desk and began to look through the books on avalanches I had not yet read. Much of the perceived treachery of snow lies in the fact that it appears stable and even welcoming, but can collapse dramatically into a lethal torrent. There are several types of avalanche, including the sort that caused havoc in Lewes where a cornice breaks from a high ridge. To the mountain-dweller or skier the most dangerous is the slab avalanche, which occurs when a raft of snow detaches itself from the snow beneath and slides downwards, gathering speed. The momentum of this surging, foaming mass can be tremendous. Avalanches have been known to smash stone houses on one side of a valley, rush up the other side to destroy some more, then return like a pendulum to finish off the first village.

The largest avalanches can accumulate millions of tonnes of snow and debris and travel ten kilometres, demolishing towns in their path, and killing thousands of people. They can knock trains off tracks, roll bulldozers over, and even push houses down hillsides while their occupants sit inside watching television. Some avalanches are small and benign, but because snow exists on mountains in such great quantities, avalanches are potentially lethal. As Philippe said, sometimes you don't even hear it before it hits you.

One of the surprising things about avalanches is how frequently they occur. The father of avalanche science, a Swiss federal inspector of forests, Johann Wilhelm Fortunat Coaz, estimated in 1910 that there were 9,368 avalanche tracks that could affect populations in the Swiss Alps alone, and between them they produced 17,480 avalanches every year. More recently, it has been estimated that there are 100,000 avalanches annually in the United States, and even this is thought to be a conservative figure. Relatively few people are killed by them in North America simply because the mountain regions are less densely populated than the Alps, but this is changing, and as the appetite for winter sports and snow-machining increases, the number of avalanche deaths per year is rising rapidly.

The first reliable account of avalanches was recorded in the first century AD by the Greek traveller and geographer Strabo, who appears to have had a rudimentary understanding of the process by which slabs of snow can come to be released. Strabo was touring the Alps, where he found that 'If one made even a slight misstep out of the road, the peril was one from which there was no escape, since the fall reached to chasms abysmal. And at some places the road there is so narrow that it brings dizziness to all who travel it afoot.

'Accordingly, these places are beyond remedy; and so are the layers of ice that slide down from above – enormous layers, capable of intercepting a whole caravan or of thrusting them all together into the chasms that yawn below; for there are numerous layers resting one upon another, because there are congelations upon congelations of snow that have become ice-like, and the congelations that are on the surface are from time to time easily released from those beneath before they are completely dissolved in the rays of the sun.'

Hannibal is thought to have suffered enormous casualties from falling snow when he crossed the Alps with his army in 218 BC: the toll for the entire crossing, including those lost to native tribesmen, avalanches and falling over cliffs, was 18,000 men, 2,000 horses and several elephants. The poet Silius Italicus wrote 250 years later in his poem *Punici*: 'There, where the path is intercepted by a glistening slope, [Hannibal] pierces the ice with his lance. Detached snow drags the men into the abyss and snow falling rapidly from the high summits engulfs the living squadrons.'

Even the most experienced guides could not protect travellers from avalanches. In December 1128, Abbot Rudolf from Saint-Trond, which is near Liège, led a pilgrimage to Rome over the Great St Bernard Pass. 'As though fixed in the jaws of death we remained in peril by night and by day,' he wrote. The threat of avalanches meant that the pilgrims were terrified of staying and terrified of proceeding. In the village of Saint-Rhémy where they had stopped, their guides, or *marones*, as they were called, refused to lead the pilgrims on. 'The small village was overcrowded by the throng of pilgrims,' continued Rudolf. 'Huge masses of snow often fell from the lofty and rugged heights above, carrying away everything in their path. Those pilgrims that were still waiting near the houses to be accommodated were swept away. Other guests, who had found accommodation, were nevertheless crippled by the fall. In such a continual state of death we spent several days in the village.' Finally, the *marones* were offered such a handsome reward that they were tempted to open the track. They wrapped themselves in felt, put rough mittens on and pulled on high spiked boots while the travellers went into the church to pray. Soon, Rudolf continues, 'a most sorrowful lament sounded through the village, for, as the guides were advancing out of the village in one another's steps, an enormous mass of snow like a mountain slipped from the rocks and carried them away... Those who had foreseen this disaster dashed to the murderous spot and, having dug the men out, were carrying them back. Some were quite lifeless. Some, half-dead, they carried upon poles, while others they dragged by their broken limbs.' When the pilgrims came out of the church and saw what had happened, they fled back to where they had come from.

Perhaps the most horrific story of avalanche survival dates back to 1755, when the village of Bergemoletto in the Italian Alps was all but wiped out on 19 March by a succession of snow slides that fell from the mountains. Three women, Mary Anne Roccia Bruno, in her mid-forties, and her sisters-in-law, Anne Roccia, in her twenties, and eleven-year-old Margaret Roccia, were buried.

They had been in the stable with Mary Anne's six-year-old son Anthony, along with six goats, an ass and five or six hens, when Mary Anne saw a mass of snow breaking down towards the east, and hurried into the stable and shut the door. They heard the roof collapse over their heads, and when the tumultuous smashing and splintering was over they discovered that they were trapped in a tiny space little more than three feet wide. Snow covered the building to a depth of forty-two feet above the roof. Several animals in the stable had been killed and their corpses soon began to stink, but two goats remained alive, one of which was pregnant. The four survived for some time on the first goat's milk, and after that died the second goat gave birth. They ate the kid and drank its mother's milk. On the sixth day, Mary Anne's son Anthony became ill. He lived on for six days until, in the cruellest moment of the ordeal, he died in his mother's lap.

Joseph Roccia, fifty, and his fifteen-year-old son had been on the roof of their house clearing the snow that had fallen unceasingly for three days when they saw the avalanche and only just escaped in time. It was five days before Joseph recovered enough from his experience to be able to dig, with his son and two brothers-in-law, but though they delved deep in many places, they could not find the house. Weeks passed, and eventually the snow began to melt. They finally found the house, but it was empty, and the search was called off until the night that Mary Anne's brother, Anthony Bruno, dreamt that his sister begged him to come and look for her in the snow. They began to dig again. On 24 April, thirty-seven days after the avalanches, they found the stable and its emaciated occupants. The three women had been entombed with the dead animals and, for much of the time, with Mary Anne's dead son. They were covered in their own faeces and urine, 'unable to walk, and so wasted that they appeared like mere shadows'. When pulled from the wreckage, they couldn't stand on their feet, and 'looked more like dead carcasses than living bodies', according to one contemporary account. The interview Mary Anne gave to an Italian doctor afterwards is cited as the first documentary evidence of post-traumatic stress disorder.

But the most lethal series of avalanches in history occurred during the Tyrolean campaign of the First World War, when tens of thousands of soldiers were killed as a result of deliberate attempts to trigger avalanches above enemy positions. In his book *Battle over the Glaciers*, Walter Schmidkunz, who took part in the campaign, wrote: 'The White Death, thirsting for blood, claimed countless victims in the mountains. Whole barracks filled with happy men, dashing patrols and marching columns, were buried in the raging avalanches that followed the blizzards. Hundreds upon hundreds were the men gripped by the white strangler. Here and there some were quickly rescued, while others remained for a terror-filled day with both feet in the grave. But these were rare occasions. The snowy torrents are like the deep sea; they seldom return their victims alive. The bravest of the brave are covered by the heavy winding sheet of the avalanche. It is no glorious death at the hands of the enemy; I have seen the corpses. It is a pitiful way to die, a comfortless suffocation in an evil element, an ignominious extinction for the Fatherland.'

One sunny afternoon in the Wildstrubel, we had played hide-and-seek with our avalanche transceivers, Philippe burying one a foot deep in the snow while Nick and I stood some distance away with our backs turned. When Philippe gave the word, we would switch our transceivers from 'transmit' to 'receive' and try to locate the beacon in the shortest time, working slowly and methodically inwards along the curved loops of the signals generated by the buried device. After an hour's practice we became efficient at finding the 'victim'. Even so, I realized that with the average life expectancy of an avalanche victim being just fifteen minutes, the best way of surviving was to avoid it in the first place.

Writing in the sixteenth century, the Alpine traveller Josias Simler described a number of snow survival techniques that seem surprisingly modern, from the phenomenon of people roping themselves together to cross snow-covered glaciers, to the use of shoes made of flat circular pieces of wood, like the tops and bottoms of wine casks, to walk through deep, powdery snow. Simler also revealed some knowledge of avalanche behaviour. He described how the snowfall could be precipitated by the smallest disturbance – the flight of a bird overhead, the voice of a passing traveller – and sweep away trees, houses, men and beasts, and carry them to the very base of the mountains. 'This sort of snow often covers an area of several acres, and falls with such a crash that the very earth seems to be shaken, and people dwelling far away, who have no knowledge of what is happening, think that they hear the sound of thunder.' The most likely spots for avalanches were known to the people of the Alps, while travellers would try to reduce the risk by setting out early, before the sun was up, and hurrying past the most perilous places in silence.

Simler had come across many accounts of men surviving, having been buried alive beneath avalanches. 'If a man is able to move his hands about, and clear a little space around him before the snow has hardened, he acquires a certain power of breathing underneath the snow, and may keep alive for two or even for three days,' he wrote.

The scientific study of avalanches and how to avoid them began in earnest at the end of the nineteenth century. In 1872, the Swiss inspector of forests, Wilhelm Coaz, invited his colleagues in other cantons to co-operate in a nation-wide avalanche survey. By 1878, statistics were being compiled in all the Alpine cantons and in 1881 Coaz produced a hefty volume titled *Avalanches of the Swiss Alps* which included revolutionary ideas on the origins of avalanches, the amount of damage they caused, and possible countermeasures. The most useful weapon against avalanches is knowing where they are likely to fall.

In 1936 the Swiss Federal Institute for Snow and Avalanche Research was created in Davos. By collecting data intensively, this first avalanche research institute in the world helped to make avalanches more predictable and offered practical advice on where to place buildings and how to design them. The terrible winter of 1950/51, in which ninety-eight people were killed in 1,300 destructive avalanches in Switzerland, emphasized the importance of the institute's work. With a programme of building anti-avalanche barriers (of which there were 20 kilometres in 1951 and 400 kilometres in 1999), the closure of dangerous roads and the identification of run-out zones, the threat has been greatly reduced, although the pressure of development in the valleys has created some new problems.

Avalanche science now plays a major role in protecting skiers. Most alpine resorts have avalanche specialists who assess risk and advise the ski patrols which runs to close, and when and where to artificially trigger avalanches. In areas of the world with extremely high snowfalls, such as Alaska, avalanche experts are employed to advise transportation departments when to close roads and when it is safe for trains to travel.

We are not yet masters of the avalanche, but neither are we still burying eggs at the foot of infamous slopes in the hope that they will ward off the evil spirits.

After our visit to Lewes, I started thinking about snow again, and decided to continue my journey. If anyone could teach me how to live safely among the mountains in winter, it was a ski guide named Matt, one of the great snow survivors. He had skied for decades in one of the most dangerous avalanche areas in the world, alone for much of the time and beyond the reach of any rescue service. That winter, I set out to meet him, in the Chugach Mountains of Alaska.

CHAPTER 7

Territory of the Mind

Valdez, Thompson Pass

It was only because of Al, the fisherman, that I managed to catch the flight to Valdez that day. Bad weather had trapped us at Anchorage airport for a morning, and we fell into conversation in the white-walled lounge next to the departure gate. It turned out that Al was a friend of the pilot of the plane we were waiting for, and he kept phoning Al on his mobile to update him on the latest weather conditions in Valdez. So when the pilot decided that the cloud was too low, Al knew even before the check-in clerk, and got me a place at the head of the standby queue for the evening flight.

Al was forty-three, with iron-grey hair and a neat beard, and words just tumbled out of him. He was returning from the Superbowl in Pittsburgh, which he had attended with his dad and two of his brothers, and the bill for the whole trip had come to $23,000. The Seahawks had lost, but it was still a hell of an experience and he wouldn't have missed it for anything.

He was fourth-generation Valdez, which is about as old as families there get. Two decades after the 1898 Gold Rush had brought the first settlers, when there was still a whiff of prospecting fever in the air, his great-grandfather had moved out here. The family still had a claim up by one of the glaciers in the fjord. When Al was a kid in the 1970s they would climb into the mountains to check out the mines. One of them was still occupied by an old man who left for the south each winter with two heavy suitcases, which they naturally assumed were filled with nuggets. One summer the miner didn't return and rumour had it that he'd either banked his fortune and left Alaska for good, or else he'd died. When Al went back up to the cabin some years later, it was exactly as the old man had left it. There was still a kettle on the cold stove.

Al was a purse-seiner, fishing for salmon and halibut in Prince William Sound. In winter, when the inshore waters weren't profitable, he would go to the Bering Sea on a company boat to catch snow crab and king crab, an occupation the Alaskan government describes in its notes for employees as 'one of the most hazardous jobs in America'. The crabbing boats stayed out for months at a time in a sea where hundred-mile-an-hour winds and fifty-foot waves were common, and where mists rolled in unexpectedly, hiding ice floes that would sink a boat. Spray would freeze to everything above the waterline – handrails, booms and fishing gear, and the wheelhouse – so the crew had to break it off with sledgehammers before it rolled the crabber right under. When the spray was flying and the weather was at its coldest, two of the crew were detailed to do it in round-the-clock shifts. They sailed in these hostile waters because crabs liked the ice pack and stayed near it, so the boats were constantly on the lookout. Sometimes the crab pots would get caught by the drifting floes and dragged

miles from the fishing ground. When they found these tangled pots the crew
would jump down on to the ice to free them.

In Valdez, Al told me, the snow comes down in buckets. The flakes were the
size of an apple, and they could easily get three or four feet overnight. Once,
as Al left for Seattle, it was raining and the valley floor was covered with grass.
When he returned six days later the snow was so deep that you could barely see
the pick-up truck in the drive. Almost every year a couple of boats would sink
in the harbour under the weight of snow.

I was sorry when, after a couple of hours, Al left to join up with his pilot
friend and head into Anchorage.

Although I had come to see Matt and learn some of his techniques for
surviving among the Chugach avalanches, in travelling to this part of the world
I was also closing a circle that had begun one evening in the 1970s when I picked
a copy of Jack London's short novel *The Call of the Wild* off the school book-
shelves. Back then I had been gripped by London's northern world, what one
early critic described as his 'territory of the mind', fitting his wild descriptions
on to the country I was growing up in, mentally filling the folds of that tame,
green corner of England with leafless thickets and great quantities of snow that
extended beyond the horizon in every direction. In this world – half Jack
London's, half mine – it was eternally dusk. The low sun cast long shadows
over the snow, and the gentle English rivers where I had splashed grew a glassy
crust and became roads, along which travelled nineteenth-century characters
with names that showed they had come from all over the globe: François,
Perrault, Mercedes, Thornton, Buck, Sol-leks, Spitz.

The Alaskans aboard the tiny Valdez plane brought to mind the people who had criss-crossed Alaska and the Yukon in London's time. The De Havilland was three-quarters filled with men I took to be roughnecks, today's equivalent of prospectors, with checked shirts and big boots, baseball caps, cheeks worn by sun or wind or drink. Among them was a sprinkling of what Matt would later describe to me as 'greenies'. Opposite me sat a thirty-something couple, athletic and slim, he with a close-cropped beard, she with straight black hair, perhaps returning to one of the state's many wildlife conservation projects. Environment and oil, green and black, rub shoulders in Alaska.

The flight attendant took obvious care over who sat by the emergency exits, interviewing each of us to make sure we knew what to do should the need arise. Then the pilot opened the throttles wide and as we bundled down the runway the plane was racked by a series of oscillations that I felt must surely shake the aircraft into a jumble of bolts and rivets. As we pitched up into the evening sky and eastwards towards the Chugach Mountains, the overhead lockers, floor and windows shook in strict rotation.

I tried to distract myself from the plane's apparent death throes by flicking through the in-flight magazine, *Alaska Business*. Its editorial made the case for exploiting the oil and gas fields in the nature reserve in the north, lambasting the environmentalist 'dictators' for trying to block the energy companies: they didn't know what they were talking about, the columnist said – there were four and a half million acres of wilderness out there. Oil had been big business in Valdez ever since the 1960s, when it became the site of the Alyeska trans-Alaska oil pipeline's southern terminal, a decision which led indirectly to the 1989

Exxon Valdez oil spill, when ten million gallons of unrefined crude leaked from
a tanker into the delicate ecosystem of Prince William Sound, killing up to half
a million sea birds. In 1994 Exxon-Mobil were fined $5 billion in punitive
damages by an Anchorage jury. Two decades later, following Exxon's several
appeals, the matter was still tied up in the court system.

The cycle of vibrations relaxed as the pilot slowed the engines at the apogee
of the flight. For a few minutes we dipped and wobbled in turbulence, then
drifted down through thick cloud which hung low over the sound and the snow-
covered runway.

Inside the terminal, I hurried over to Valdez U Drive, the only car hire
company in town, which I had been trying to contact on and off for days. They
never answered, so I had left a series of increasingly desperate messages. I
needn't have worried. 'I recognize that accent,' said the woman at the counter,
addressing me by name. She gave me a set of keys and I grabbed my bag from
the luggage reclaim belt next to a sign that read 'Please Pick Up Checked
Firearms at the Ticket Counter' and wheeled it out into the snow.

The first thing that hit me was the dampness in the Valdez air; it was saturated, like arriving in the tropics, only colder. The temperature was just above
freezing and everything was melting, but the drive into town from the airstrip
through the half-light was an epiphany. A gentle, wet snow fell from a hanging
mist that divided the world into two: below, it was already night and lamps
blinked over the black water of the fjord, while above, the sunset played its
miraculous slide show on the gleaming peaks, their mood changing from yellow
to orange to purple. At the roadside, snow stood a couple of feet high.

I parked and checked into my hotel before setting out to walk through the little town, treading carefully through the ice-slush along the harbour front. Snow was piled high around the entrance to the Wells Fargo office. On Pioneer Drive, men were shovelling snow from a flat roof. This was the new Valdez, laid out in wide streets of boxy, functional architecture, interspersed with industrial structures built by the oil companies who had moved in with the pipeline. The old town at the eastern end of the fjord had been destroyed by the Good Friday earthquake of 1964. The quake had triggered an underwater landslip which had sent a tsunami charging across the fjord and into old Valdez, killing at least thirty people. The community had started again on safer ground on the north shore.

Snowploughs roared past the twenty-foot stacks of snow, darkened with grit and mud, that were dotted all over the town. A sign next to one read: 'Snow Storage Area: Danger. Do Not Enter.' These dumps held so much snow they would not melt away until midsummer.

Shouted snatches of conversation reached me from the neon-signed Pipeline Club. Drinking, I discovered, was a traditional Valdez pastime. Al had told me the Harbour Club on the front closed at 7 a.m., only to open again an hour later. Some old guys were often there until closing time – you wouldn't believe they could do it. Next morning two of them wandered into the hotel for breakfast at 7.30 a.m., with alcoholic eyes and red faces, shouting and knocking things over before they departed at the waitress's insistence.

On that first night Valdez felt like a Wild West frontier town, on the outskirts of civilization. I liked that. Some day people would step off the plane

to find a series of interconnected, centrally heated malls and bowling alleys –
but not just yet.

Returning by the harbour I heard a splashing in the blackness below.
Something was moving on a pontoon in the near dark: a creature the size of
a dog, lounging and scratching itself as though it was bathing in bright
sunshine. I saw another, in the water, swimming on its back, flippers folded
neatly on its chest. It was a family of sea otters. I stood and watched them play
between the moored boats and snow-covered pontoons, then turned back to
my hotel.

Valdez has always attracted people seeking a fast buck and adventure, egged
on by those who sold them their passage. But the extraordinary weather and
difficult country for a long time made the fjord more dangerous than lucrative.
When the first gold stampeders landed in 1898, they were surprised to find no
town here at all, just a collection of tents. They had been led to believe that
Valdez offered an easy, all-American route to the Alaskan interior and the
Klondike, notably by the Pacific Steam Whaling Company which had a cannery
in Valdez and northbound ships to fill. The *San Francisco Chronicle* and the
Seattle Post-Intelligencer made it their duty to wind the excitement about gold
strikes in the north into a mass hysteria – the *Chronicle* even published a map
showing goldfields in the Copper River just east of Valdez. The map was a
complete fabrication, as were the goldfields and the route to the interior. Instead
the prospectors found they had to cross the mountains to the 'gold' via the

Valdez Glacier, a thirty-mile-long river of crevassed ice whose 4,800-foot summit was swept by gale-force winds and avalanches, and which received up to a thousand inches of snow each winter. Even down by the ocean the snow fell in quantities none of the new arrivals had ever before seen. When the entrepreneurs of the Connecticut and Alaska Trading Company arrived in March 1898 they found the beach buried under ten feet of snow, which they had to dig out and haul away before they could start to build their log cabins. Early photographs of Valdez show people moving about in trenches and tunnels.

On 1 April 1898, the *Klondike News* published the truth: 'We warn our readers against any attempt to reach the Klondike Country by way of [Valdez and] the Copper River. No living man ever made this trip, and the bones of many a prospector whiten the way... there are trackless mountains to cross, by the side of which the Chilcoot trail [in southern Alaska] is a boulevard... Certain unscrupulous parties operating steamboats up that way are issuing gaudy pamphlets with nicely worded directions of how to travel over a country that white man has never set foot in. This is worse than murder, and such parties deserve to be punished to the full extent of the law.'

The warning came too late for the four thousand who had already landed. Equipped with man-hauled sleds and sacks of flour, crampons that looked like instruments of torture and maps so thoroughly wrong that they marked the Valdez Glacier running east–west instead of north–south, the stampeders set off into a snowbound hell.

The prospector Neal Benedict wrote of the glacier: 'Danger lurks at every hand from fathomless crevasses and sink-holes, often concealed by freshly-fallen snow, which treacherously invited the careless step, the result of which could only be a sure, swift, terrible death.' Addison Powell, who joined the Gold Rush to the Valdez and Copper River area as a US deputy surveyor in 1897, wrote of miners discussing the glacier 'smoking' – the wind whipping snow off it – and reported that even when this was happening, inexperienced or desperate prospectors, including a woman dragging her sick husband on a sled, tried to make the crossing of it. Others wrote of regular avalanches. 'Snowslides frequently occur from the surrounding mountains,' noted the prospector and diarist Joseph Bourke, from Brooklyn. 'The attention is attracted to the slide by the terrific noise which much resembles the sound made by a train of cars running into an enclosed depot.'

William E. Treloar recalled a storm at the end of April 1898: 'It had begun to snow and a storm set in in earnest. The snowflakes [were] so large and came down so thick we could hardly see our neighbouring tents only a few feet from ours... It snowed and snowed and we shovelled and shovelled till we were throwing snow ten feet high so we finally raised our tent. Nights we would build up piles of sacks and boxes under the ridge-pole to keep the snow from breaking our tent. In the morning we would have to shovel our way out of the tent then clean the snow off.' On 30 April, after the storm had dropped eight feet of snow on the glacier, an avalanche swept through the camp below the summit, burying it in snow up to fourteen feet deep. 'Half clad and barefooted [the surviving prospectors] grabbed shovels and rushed out into the cold night air working frantically to locate and free those buried by their tents,' wrote Treloar.

'The swift action succeeded in rescuing twenty-five.' At least five men died in the avalanches that day. For weeks afterwards, miners could be seen probing for their supplies, which had been buried by the snow.

Even now, people who venture on to the glacier in summer find possessions lost by the prospectors in 1898. The following day in the town museum I saw a glass case holding a number of orphaned essentials of mining life: a harmonica, which had rusted and fallen apart, a circular metal canteen, a gaming token and an iron runner from a sled. No one knows what became of their owners, whether they died on their mad mission into the interior, or whether they crossed the glacier successfully, and even found a few nuggets of gold.

'Have you much experience of driving in snow?'

As he gave me directions to his house over the phone I didn't tell Matt I was from a city that had pretty much shut down the last time four inches fell.

It didn't sound easy to reach, but at least I wouldn't get lost. There is only one road out of Valdez, the Richardson Highway, which climbs over Thompson Pass, a 2,700-foot boundary between the maritime snow of the coast and the continental snow of the interior. They closed the weather station in 1973, but even relying on pre-1973 data, it holds seven of the ten records in a table of snowfall extremes compiled by the National Climate Data Center in 2005.

I knew that Matt had been skiing in and around Thompson Pass for almost thirty years, in an upper world riven with crevasses, sluiced by avalanches, trapped with melting ice and unpredictably blasted by gale-force winds at freezing temperatures. The Chugach peaks scrape the moisture out of jet streams that have been travelling for hundreds of miles across the Pacific, and at these latitudes the storms are regular and often severe. Valdez's snow is famous enough for the town to support five heli-skiing operations, though it has no resort, no ski shops, no lifts and no pistes. The amount of snow, combined with the absence of patrols and rescue services, makes the pursuit of winter sports here a risky activity. Even keeping the pass open is dangerous. During the winter, a team of drivers ploughs the road round the clock and drops avalanche bombs into the snowpack to trigger safe snow-slides. They wear avalanche beacons so that they can be located if their machines get buried.

Matt was out on the snow 140 days a year, forty days of that alone. If he got into trouble, there was no one he could call on for help. Sometimes, if his wife Tabitha suspected something was wrong, she would drive up to the pass and see if he was in range of the radio, but it was not much protection. He reckoned he had climbed seven million feet in his time. He was perhaps the most experienced back-country skier in Alaska.

I drove up towards the pass the following day, past signs that read: 'Avalanche Area for 7.5 Miles. Do Not Stop.' There was fresh snow on the highway, and even though the ploughs had taken most of it off, a residual coating of corduroy ice seemed to have bonded with the tarmac. I tried braking hard at a low speed to see what would happen, and the results weren't good, so I slowed even more. With banks of mist rolling in and out of the woods on either side of the road, it was difficult to make out the mountains close by, and though I knew I was passing between the 300-foot-high cliffs that form Keystone Canyon, I could see nothing except the lower reaches of the rock and the waterfalls which had set into great cascading ice sculptures.

The radio was tuned to the local station KVAK FM, on which two people spent half an hour discussing the school's budget in serious northern tones, but beyond the canyon KVAK disappeared in static and I searched the frequencies. Medium wave produced National Public Radio from Prince William Sound, on which a dead-voiced DJ read out the classified advertisements: 'Yamaha snow-clearer. Almost new. $200. Call 87987...' Soon even that vanished, and the dial was empty from end to end. My mobile phone had lost its signal long ago. I was now out of range of the twenty-first century. It was just the car and me. I kept the static on anyway, searching from time to time in case something turned up.

The highway crossed a creek and climbed into low cloud, which reduced visibility to a few feet, but I made out the turning that led to Matt's house and pulled into it, stopping as the car sank to its axles in snow. In the cloud I couldn't see where the route went, or indeed if there was one; ahead there was only whiteout. Then, through a gap in the mist, I saw a track that would qualify as a piste in most ski resorts, with white walls where a plough had cut a route through. I crept towards it in low gear, and then I was in it and rolling downhill and there was no turning back.

I sped down and right and sharp left, along past a car just visible in a great mound of snow. After a quarter of a mile the track levelled out and ended in front of a small wooden house. I parked between some trees, wondering how

I would ever get back to the highway again. From within the cabin drifted the rising and falling whine of a vacuum cleaner. I called out a greeting, and a blonde woman appeared.

'Hi, I'm Tabitha,' she said. 'Matt's on his way over. Here he comes now.'

I turned to see Matt striding out of the mist between the trees in dirty padded trousers and ski jacket – all black so that they absorbed the heat of the sun, he would later tell me. His Nordic ski hat was murky with what looked like seal oil, though it must have simply been a decade's worth of sunblock, which gave him the air of an early polar explorer. He was slim, with unkempt teeth and hair, the skin around his eyes wrinkled from years of squinting into the low sun, his irises grey and clear. I guessed he was in his late fifties. He introduced himself in a slow, Midwestern voice that contained a hint of irritation, as though I had caught him on his day off, but the edge in his voice disappeared when we began to talk about snow. We should go out on the pass, he said. We loaded some skis and snowshoes into his huge all-American pickup, with its massive blade on the front, and set off up the track towards the highway, ploughing as we went. As he drove, he told me how he had come to live in this remote country.

Matt had left his native Oklahoma in 1975 because he was partying more than was good for him. He joined the US Coast Guard and was made a navigator. His first posting was to the Bering Sea, where he served for four years, until the day he saw an aerial photograph of Valdez in winter, all buried under snow, and decided that beautiful and tranquil place was where he wanted to be. He arrived in the autumn of 1979, and bought the six acres of land near the pass, where he now lives. It wasn't until the mid-1980s that he started work on a small log cabin, stripping the timbers himself. It took him eight years, off and on, to build the tiny, down-home pioneer's house with a sleeping platform in the eaves and a rickety spiral staircase made out of small branches.

As we climbed, the sun broke through the cloud. It was turning into a clear, sparkling day, with a wind that blew the snow in wisps down the road and silted up the edges. From the snowbanks on either side, the drifting grains were rising, forming twisting patterns like smoke in a wind tunnel. The eastern Chugach came into view, through the Marshall Pass, and scanning the horizon westwards towards Valdez I saw a series of peaks, all cast in pure whiteness. We passed the summit and a sign informing us that we were at 2,700 feet, together with an anemometer and, further on, the Department of Transportation base tucked behind the hill, then drove on into the pass itself, a wide-bottomed glen surrounded by peaks up to 7,000 feet high, with views down the valley to the north and the interior. We parked and climbed down from the truck, inhaling the cold mineral smell of snow. Matt put on a pair of skis with skins fitted underneath; I pulled on my snowshoes, and we set off uphill towards a group of National Park huts, which were buried to their rafters. There wasn't a human or animal track in sight. We walked and slid to the foot of the Worthington Glacier, and up along its northern edge, then turned to look back at the barrel-shaped valley, Matt pointing out where the avalanches most frequently occurred, and where he had seen people get into trouble.

Matt's method for staying alive came down to his knowledge of the local area and the snow, combined with a caution and patience that most people would have had difficulty sustaining. The technique that protected him most of all was his constant measuring of the slope angles with an inclinometer. He was a follower of the Alaskan expert Jill Fredston when it came to avalanches. By knowing the steepness of the slope, by 'putting numbers on it', you could eliminate much of your uncertainty about whether the snow would slide, he said. Slab avalanches, the type most off-piste and back-country skiers get caught in, can reach speeds of up to 150mph. Almost all of these occur on slopes angled between 30 and 50 degrees, with the most dangerous angles being between 35 and 40 degrees. Above 60 degrees, the snow tends not to stay on the slope in the first place; below 25 degrees, the snowpack will tend not to slide, however unstable it is. So the most important tool you can carry, said Matt, is the one that measures the angle of the slope. It gives you a solid number, rather than relying on intuition. It is something concrete on which to base your decision.

We climbed along the northern side of the glacier, trekking S-bends up the slope, until he stopped again and pointed to the packed snow we were walking across. This was wind slab, he said, just the kind that could make an avalanche happen. Snow crystals are fragile, and when they fall the wind can break the little prongs off, or bounce them along the ground so they become rounded and pack together more tightly. The wind can also press down on the snow. In this way it causes layers of dense snow to form within the pack, and these sometimes don't bond well to the snow above or below. Matt carved a solid chunk out of the slab with his ski stick and tested it on his knee. If it is hard and doesn't break it would probably support a skier, he explained, but if it breaks easily it would also break with someone on it. The wispy patterns on the slope to our left showed where the wind had removed the snow to deposit it where we were standing. Jill Fredston describes these as arrows that point in the direction the snow has been carried.

A little further up, Matt unpacked his shovel and started digging. There are many types of avalanche pits you can dig, he said. For some tests you cut a really big block out of the snow and stand on top of it with your skis on and see if it breaks. If it doesn't crack, you jump up and down on it to see when it will. Then there are scientific pits where you use a thermometer to measure the temperatures of the different layers of snow at successive depths. If the temperature gradient is small it's probably okay. If there's a substantial difference between one part of the pack and another, that means it's very unstable. There were days when Matt had skied on a slope, gone home to his cabin and noticed a temperature change on his thermometers. When he came out the next day the whole hillside was lying at the bottom of the valley.

With his shovel, Matt carved a chunk out of the slope about three feet deep at the back wall. He took off his glove and, starting from the top crust and working in a vertical line towards the bottom of the pit, pressed his finger gently into the snow every half-inch. His finger made a column of similar-sized indentations for a foot, after which the indentations stopped, and four inches further they started again, and continued to the bottom. This indicated that between the two softer layers there was a shelf of hard, dense snow. The question now was whether the top, softer layer had bonded to this harder layer beneath it, or whether it would slide over it. He began to cut out a column ten inches across with the shovel. When he picked it up, it broke into three distinct layers, like chunks of polystyrene, as the two softer outer layers separated from the hard centre. It had rained and snowed together on the pass over the past few days, Matt said, for around six to nine inches, then that had frozen, and a foot of fresh stuff had fallen which hadn't bonded to the layer below. He looked up at the top of the slope above us which, to my eye, was not especially steep. 'That's why we won't be going up there,' he said. 'Might not look much but there are tons of snow on that hill and if it starts to slide it could kill us both.'

Matt knew the terrain could kill him at any moment if he made a mistake (and it could quite possibly kill him even if he didn't). There are few certainties when it comes to avalanches. He also knew he wanted to live to ski the next day. So he made the odds as favourable for himself as possible, sticking to the low-incline slopes wherever he could. He gave himself the best chance.

In all his three decades of unprotected skiing he had never fallen into a crevasse or been caught in one of the many avalanches that sweep the pass. He'd had a few close calls, though, like the time he'd broken his leg while he was out alone, and had to ski a few thousand feet on one leg, with his snapped bone-ends grinding against each other whenever he jarred it on the way down. He'd managed to run into the only patch of brush on the whole slope. Having crawled through that, he had reached the road and tried to flag down each of the few cars that passed, waving at them to stop, but they had just waved back. Eventually, a friend had pulled over and taken him as far as the nearest phone. Matt had to make his own way to the hospital because his friend was going to Anchorage and didn't want to be late.

Then there was the time he had seen a beaver dam and decided it would be a good idea to ski across it. He was right in the middle of it when he put his ski through the ice and into the water up to his thigh, and he couldn't free it no matter how hard he tried. The binding was stuck, so he couldn't just remove his ski. It had been about −20F (−28C) that day, but he forced himself to take his jacket and shirt off, strip to the waist and bend down in the water with one leg in its ski still up on top of the dam. After a while he had managed to wriggle his boot free and put his clothes back on before he froze, then he'd fought his way through the chest-deep snow with only one ski, back to the road where his car was parked. When he finally arrived home he'd slept for eighteen hours solid, exhausted by the adrenalin that had been pumping through him.

Sometimes, Matt said, he thought he'd like to retire to work as a lift operator in some Colorado resort where he'd become known as that crazy old guy who says he used to ski at Valdez. I couldn't really see him working the lifts in Aspen, not when he'd built his own house in a place where the snow lasted six months of the year. Out here, he had developed his own eco-friendly, lift-free version of wilderness skiing. He had done this partly by force of circumstance – a plan to build ski-lifts in Valdez had foundered in the 1980s – but also because he had come to believe it was right. The tranquil white wilderness he had seen in the photograph that had drawn him to this place was better without the clanking, humming paraphernalia of a lift system. He railed against the flashy heli-skiing operations at Valdez which landed customers on the tops of the mountains around the pass, sometimes directly above him. 'Goddamn movie stars,' he called them. 'Always trying to kill me.'

Back at the cabin, I left Matt and Tabitha clearing a path for visitors who were arriving later in the day. There were animal footprints in front of the house, where moose had come to eat the shoots that were sprouting from the trees. The sun was streaming warm and strong through the wood, loosening clumps of ice crystals which dropped silently into the snow that already covered the earth. 'Looks almost like spring has arrived,' Tabitha said. I couldn't tell whether she thought this was a good or a bad thing.

As I brushed the snow off the car and defrosted it, I thought about Matt and Tabitha's life in that remote place beyond the city limits. Among the moose and ptarmigan and bears, it seemed to me they were not too far from the stripped-down living experiment Henry David Thoreau had embarked upon for two years in the 1840s by moving into a cabin next to Walden Pond, on Ralph Waldo Emerson's land. 'I went to the woods because I wished to live deliberately, to front only the essential facts of life, and see if I could not learn what it had to teach, and not, when I came to die, discover that I had not lived,' wrote Thoreau. Matt and Tabitha similarly had created a home on an empty patch of land with very little but the resources they discovered there. Matt had looked for a new corner of the world, somewhere he could settle and live freely, and he had found it.

John Updike once remarked of Thoreau: 'To the dark immensity of Nature's indifference we can oppose only the brief light, like a lamp in the cabin of our consciousness; the invigorating benison of Walden is to make us feel that the contest is equal, and fair.' Matt made me believe that the contest with the snow-clad mountains could be fairly balanced; that I too had a chance.

I took a run at the drive, and made it to the highway in one go.

✳

In my hotel room that night I pored over my well-thumbed copy of Jack London's collected stories. London, then a young, unpublished writer, had spent the pitiless winter of 1897/8 in the Yukon, prospecting for gold but also collecting the experiences of others who had raced to the Klondike to get rich. In the procession of northern stories that flowed from this adventure he described in detail what he had seen, recounting the way in which starving wolves would follow dogs and men for days or weeks, picking them off in the night when they were at their most vulnerable, how people froze from the extremities inwards, feet and hands following toes and fingers into numbness, before the victim felt an illusion of warmth and then a fatal urge to sleep. In outline, the stories may have been those he'd gathered from people he met on the trail, but the thoughts and feelings of these characters were drawn from London's own experience. There was surely much of his psyche in *The Call of the Wild*, in which the canine hero Buck is kidnapped from California and taken north, where he becomes a more savage and stronger realization of himself. This same self-fulfilment is there in London's story 'To Build a Fire', which tells of a prospector who ignores the advice of an old-timer that 'no man must travel alone in the Klondike after fifty below'. When he breaks through the ice on a day when the temperature is −75F, soaking his boot and lower leg, the tenderfoot can only save himself by building a fire, which he foolishly places under a tree. The snow in the branches thaws and extinguishes the flames. From that point he is doomed, and we follow him through his last hours, observing the cycle of hopes raised and dashed, until he lies dying in the snow with the old sourdough's words ringing in his head: '*No man must travel alone in the Klondike after fifty below.*' Arrogance killed the prospector. Small human weaknesses are fatal in the accelerated Darwinism of the Klondike winter.

Jack London's stark vision can be traced back to the economic depression of late-nineteenth-century America, and to the writer's early life. Financial hardship coloured much of his short childhood in Oakland, California. The man most likely to have been his biological father, the itinerant astrologer William Chaney, abandoned his mother, Flora Wellman, when she became pregnant with Jack. Soon after his birth, Flora married John London, an impoverished Civil War veteran with two daughters in an orphanage. When Jack learnt just a few months before he travelled to the Klondike that he was not the son of John, he wrote to Chaney, who flatly denied being his father. His disputed paternity, combined with the family's poverty, influenced London greatly. One biographer wrote that Jack saw himself as a sort of wild 'feral orphan', like Kipling's Mowgli, for much of his life. The character that emerged as a teenager was scrapping and tough, determined or cursed to live at a phenomenal pace. By the age of fourteen, he was working shifts of up to twenty hours in a cannery to support his family. At fifteen, having taught himself to sail, he became an oyster pirate, raiding the commercial beds in San Francisco Bay. By sixteen, he was a hard drinker and unhappy enough to make a suicide attempt, which he aborted at the last minute. Aged seventeen, he signed on for a seal schooner bound for the coast of Japan and the Bering Sea, and on his return joined the army of unemployed who marched on Washington to protest against the government's handling of the Depression. He deserted the demonstration in the Midwest, and stole rides on freight trains to the East Coast. In upstate New York he was arrested for vagrancy and sent to Erie County jail for thirty days, where he was probably sexually assaulted by other prisoners. He returned to Oakland when he was nineteen, convinced that education was the best way out of poverty, and embarked on a lightning programme of learning. By the time he reached the Klondike, aged twenty-one, near the front of a tide of 100,000 men and women who made up the greatest Gold Rush in history, he was an experienced traveller, a socialist and a budding writer.

In Alaska and the Yukon, London encountered a desperate combination of extreme weather and people who were unaccustomed to it: men and women who had put their life-savings into the rush to find gold, breaking down, giving up, or meeting their deaths on the trail. Pack animals froze or starved by the thousand; the carcasses of horses and mules littered the route between the coast and the interior. London crossed from south-east Alaska to Canada by the Chilkoot Pass, where his path traversed several avalanche run-out zones on its climb into the clouds. The snow here was thought to be 70–100 feet deep, and the stampeders had to make many trips through it to carry to the top the ton of gear each person required. The classic photographs of the Gold Rush were taken there. They present an enduring image of the human spirit, suffering and futility: the laden-down prospectors queue up for the two-thousand-foot climb to the top of the pass, which they then scale nose to tail in single file, black ants against the snow, bent double under the weight of the crates on their backs. When they reach the summit, they drop their loads, catch their rasping breath and turn back down to pick up the next one.

Jack London was fit and resolute – he boasted that he could carry his ton of gear two miles in a day. From the Canadian border-post at the summit of the pass, he portaged his supplies to the shores of Lake Lindeman, where he and his partners built a rough boat, crossed the lake and travelled downriver, against the onset of winter ice and snow, and through several sets of rapids. He reached Stewart River, a now-abandoned settlement where a tributary joins the Yukon, as the freeze set in, and staked a claim twenty miles away up Henderson Creek. However, most of the lucrative mines in the Klondike had long been claimed, and his produced no gold. His health broke down in the winter and he developed scurvy from living on a diet of preserved food. With the spring thaw he sailed down the Yukon and across the breadth of Alaska, reaching the sea at Saint Michael twenty-one days later. He worked his passage back to Oakland, where he discovered that his stepfather had died and that he was now the sole provider for the family.

Jack London found no gold in Alaska, but he had found his fortune. Within a year of his return to California, his first story, 'To the Man on the Trail', was published in the *Overland Monthly*. International fame came with the publication of *The Call of the Wild* in 1903. Over the next ten years he worked and drank with equal determination, and by the time of his death, aged only forty, he was the highest-paid writer in the world, with fifty published books of fiction and non-fiction, including two hundred short stories and nineteen novels. As well as travelling to the Yukon, in his short life he had crossed the Pacific several times, covered two wars, married twice, become a rancher, built his dream home only to see it burn to the ground, and written a memoir of alcoholism, *John*

Barleycorn, which became the set text of the Prohibition movement. He established literature's fascination with boxing with his story 'The Abysmal Brute' and his novel *The Game*, and with the road-trip: his book *The Road* became the inspiration for Jack Kerouac, paving the way for later generations of adventuring writers, from Orwell to Hemingway and Norman Mailer. He is still among the most widely read of American authors, and his most popular books are the vivid northern stories which have transported millions to the Klondike during the Gold Rush madness that gripped the US and Canada in 1897–8.

Staying alive is the highest achievement of Jack London's inexperienced, ill-equipped heroes. Maybe during his time in the north London's greatest achievement was his survival. In the close proximity of death, surrounded by this cold, white stillness, he became acutely aware of what living meant to him. In the Yukon winter, the man who once tried to end his life had found not only some distance with which to view his relationship with his father, but had reached an accommodation with his own existence.

The following day I set out into the interior to try to reach Copper Center, where the most dogged of the Valdez prospectors arrived and set up an encampment before winter in 1898. They had taken the route over the glacier. At least I could take the road. From Thompson Pass the highway ran north along a valley which opened out into a broad plain. I passed the occasional oil truck and snowplough along the road, but little else. After the mountains, the air felt dry and cold: snowfall here in the interior averages just 39 inches a year, but temperatures can still drop to −75F, the temperature at which, according to London, spit freezes when it leaves your mouth. The drama of the coastal landscape, where mountains plunged straight into the sea, was absent. Instead, the land, densely covered with low trees, stretched flat to the horizon under a big, empty sky.

The prospectors hadn't stayed here long. Some had returned over the glacier before the winter had set in. Of those who remained, many others succumbed to scurvy and several died. In the spring of 1899, the survivors headed back to the coast. The Gold Rush sparked off by the Pacific Steam Whaling Company was over, but it left its mark. Many of those who had bought a passage to Valdez expecting to find a town stayed on and built one. With help from the army, they even managed to cut a trail into the interior. This is the route through Keystone Canyon and over Thompson Pass that the present highway follows.

There is a still a settlement at Copper Center, with a hotel, a garage and a gas station, all tidily built from tree trunks. Al had sung its praises to me: his family had once had a gold-panning business here. I arrived desperate for food, and was glad to find some at the Copper Center Lodge. Next door there was a small exhibit to the Gold Rush, but it was closed during the winter months. I wandered up the track between the buildings. A four-wheel-drive roared along the dirt road, its driver yee-hawing before disappearing in a cloud of dust. Under this gaping sky, the human endeavour of Copper Center seemed somehow precarious.

I did not stay long. The journey being more significant to me than the destination, I drove south again listening to the only radio station I could find, which was devoted entirely to religion. It was broadcasting a browbeating sermon. Did the emptiness of the interior make people particularly mindful of their souls? 'What do you do to deal with your loneliness?' the radio evangelist asked, and I was reminded of the White Silence, a time Jack London identified in the north when the sky was clear, the heavens were as brass and the slightest whisper seemed a sacrilege. 'Strange thoughts arise unsummoned,' he wrote, 'and the mystery of all things strives for utterance. And the fear of death, of God, of the universe, comes over him – the hope of the Resurrection and the Life, the yearning for immortality, the vain striving of the imprisoned essence – it is then, if ever, man walks alone with God.'

It was late afternoon when I reached the Thompson Pass. I parked the car and climbed out into the cold. The sun was out and the mountains were spectacular. Elsewhere, mountains below a certain contour turn black and brown, even in winter, but here the mountains were so thickly iced from top to bottom that every indentation was filled with white.

Once over the other side of the pass, I pulled off the main highway and followed an unploughed side-road that led to the foot of the route the prospectors had taken across the Valdez Glacier. These were my last hours in Alaska, and I was damned if I wasn't going to walk out in the snow one last time. I rooted in my bag for clothes, put on my ski trousers and snow boots, and set off along the covered-over track. I quickly worked up a sweat in the deep snow, the only sound the grinding of my boots, the swish of my clothing and my panting breath.

The top of the scrub was visible through the surface of the snow. I heard the call of a ptarmigan, and thought about the wolverines and the moose, and didn't the bears sometimes wake from their hibernation to look for food? What else was lurking in the wilderness? After a while I came to a halt. My route over the curved land suddenly levelled out flat and I realized I was at the edge of a wide glacial lake. In the distance was the tumbling mass of the glacier, and the only way ahead lay over the water. Surely, in deep winter, the ice must be thick, but with the humidity I couldn't be certain. I put my boot on it and heard a clicking sound. Was it the trees, or was it the muffled noise of the ice under my feet, or was it in my head? I couldn't tell.

I imagined the walk in my mind's eye, the crack I would hear somewhere too far from the edge, my scrambled attempts to get free, the broken ice tilting under my weight, the impossibility of getting a handhold, the water soaking my clothes and freezing where it met the air. I thought about Matt and the beaver dam. If he could get into trouble, how easy was it for me? Nobody knew where I was. How long before they would come looking? When would Lucy know? In a week? A month? They would find the car first. Someone would become suspicious. The state trooper would be called out. KVAK FM would put out a bulletin and the news would travel rapidly around the town. In the Pipeline Club, they would talk about the English guy who had told no one he was going for a walk alone by the glacier. Damned tenderfoot, they would say. My cheeks flushed despite the cold.

Then I remembered Philippe and his admonishment on the Daubenhorn. *I think you are dead while you are alive, Mister Journalist. I know you are frightened… but you take risk unconsciously. Take this risk consciously now, with me.*

There had been a time when I was not afraid of death. I would boast to myself after my father's death that I had witnessed it close at hand and knew the empty hole it left behind. It did not hold any terror for me. I knew that to really live I needed to take the risk of not living. But somewhere along the line I seemed to have forgotten all that. How? Children, perhaps. My own life now seemed so much more precious when reflected in theirs. Perhaps I had also kept to the swaddled comfort zones of Western life for too long. Between them, Philippe and the Alps had exposed my soft centre. Now was the time to start to make amends.

I took off my gloves so that when the ice broke I could better grab on to its fractured plates. Then I stepped out on to the lake's surface.

I walked thirty careful metres on to the lake, every hair on my head alive to the first cracks, envisaging the unknown black depths beneath the white crust. I stood in the middle for a few seconds then walked back, taking the last steps deliberately slowly. My fingers felt a painful heat, but I didn't care.

When I reached the shore I was shaking with adrenalin. 'Fuck you!' I yelled as loud as I could – at the bears, the ptarmigan, the moose and all the unknown fears beyond, but mainly I shouted it at Philippe.

'It was in the Klondike I found myself,' Jack London once wrote. 'There, nobody talks. Everybody thinks. You get your perspective. I got mine.'

When I reached the car the sun had slipped behind the mountains. Only their tops were lit up in their dusky colour show. I switched on the engine and pushed the heater up as high as it would go.

CHAPTER 8

The Art of Winter

Vienna, London

Within two weeks of arriving back in the dark, damp London winter, I was itching to escape, if only for a few days. But I couldn't leave Lucy again so soon, when Eddie was still so little. What if the three of us went and Harry and Arthur stayed behind in London with their grandparents? We could go to Vienna, say, where we had never been, for a city break. It had been an exhausting few months for Lucy and she was easily lured by the promise of the city's art galleries, the Leopold Museum with its Klimts and Schieles, and the art deco Secession Building, as well as the Vermeers and Rembrandts, Van Eycks, Dürers, Caravaggios, Titians, Veroneses, Tintorettos and all the rest, scattered around various baroque palaces. I had my own reason for choosing Vienna: I wanted to see the world's most famous winter painting, Pieter Bruegel the Elder's *The Hunters in the Snow*, which hangs in the Kunsthistorisches Museum.

We boarded a flight one blustery winter morning, and a few hours later were installed in a hotel in the city's museum quarter. We then set out to explore the Austrian capital.

The façade of the Kunsthistorisches Museum is one of an Italianate matching pair that face each other across the garden of Maria-Theresien-Platz, near the former imperial palace, the Hofburg. It displays the art collection of the Hapsburg Empire, a stunning hoard of treasures put together by a dynasty that ruled much of Europe for five hundred years. I found the Bruegels in the east wing. Here a small gallery contains the most extensive collection of the artist's work in the world – a dozen paintings of varying shapes and sizes. In its simple frame *The Hunters in the Snow*, painted in 1565, hung on a wall the dull blue colour of a winter sky.

As a child I would often stop and look at this portrait of a frozen world in a reproduction my grandmother had in her house in Yorkshire: the down-trodden hunters struggling through the wind and snow towards their village, and the skating pond in the middle distance. Was this, I wondered, a scene from my grandparents' youth? I could not imagine a time longer ago than that. And where? Which mountains plunged from such high summits down to a plain of flooded fields? What did it say on the inn sign, and in which language? Now, many years later, I was standing before the painting again, only this time it was the real thing. I leant in over the velvet rope for a closer view.

The viewer stands at the top of a low hill, next to a straggly blackberry bush weighed down by ice crystals, looking across a broad valley towards a fantastical collection of Matterhorns and Eigers, whose foothills reach towards a walled city at the edge of a frozen sea. A band of huntsmen have just arrived on this epic set from stage right, and are struggling through the drifts and wind with their hungry dogs. They pick their way among the black vertical stripes of the tree trunks, carrying their meagre bag, a fox, past an inn where peasants are singeing the hair off a slaughtered pig in a fire blown by the breeze into a furnace. The branches above are laden with snow which will drop with a splat at any moment. Birds in the sky and in the trees seem to mock the hunters, as do the fresh tracks of game which cross the snow to their right. The inn sign, showing Saint Hubert, patron saint of huntsmen, hangs precariously on one hinge.

The painting was larger than I had expected – perhaps five feet across – and the detail beyond the huntsmen far clearer. Below the hill, villagers act out their lives: a man on a short ladder halfway up his roof throws snow at the chimney which has caught fire, while his neighbours hurry towards the house with ever longer ladders. On the frozen ponds in the middle distance children are spinning their wooden tops, and a man has fallen flat on his face. People are curling and playing kolf, a precursor to golf. A romantic couple hold hands while three children skate past in formation. I could almost hear their shouts drifting up the hill. A girl in an apron pulls her friend or sister to the lake on a sledge. The watermill, a symbol of industry, is frozen to a standstill.

Looking over the snow-covered land, what did Bruegel see? Not misery. He is more even-handed than that. He sees the pleasure of winter as much as the pain. These are lean days, as the huntsmen's bent backs attest, but they are also days of fun and leisure. Apart from the business of hunting and gathering wood, work has largely stopped. People have come out to enjoy themselves on this clear, special day, when the ice has made the landscape new. It is also a time for children, for innocence and play, for romance and for games. Bruegel, it seemed to me then, had created a detailed and truthful account of the world's reaction to snow.

Half a dozen people had by now gathered beside me, and not wanting to appear to be hogging the picture, I moved on around the room. *The Procession to Calvary* and *The Tower of Babel* are typical of Bruegel's work before 1565: religious allegories, transposed from biblical times to sixteenth-century Flanders. On the same wall as *The Hunters in the Snow* hang *The Gloomy Day* and *The Return of the Herd*, which Bruegel painted in 1565 and which, together with *The Hunters*, form part of a cycle of six paintings that depict the seasons of the year. This series marked a departure in style as well as subject for the artist, and would prove to be a turning point in the history of art. Why, in 1565, did Bruegel turn his attention to such an apparently mundane subject as the weather?

The painting of the seasons was not an original idea: for centuries medieval artists had illustrated the labours of the months, with peasants and nobles planting, harvesting, hunting and slaughtering, under God's watchful eye. Among these, the Limbourg Brothers' *Très Riches Heures du Duc de Berri*, painted between 1412 and 1416, is perhaps the most famous. Their painting of February shows the hillside white with new snow. Under a darkening sky a young man cuts wood while a peasant returns to the farmhouse in the foreground with a blanket over his head. Inside the house, two others warm themselves by a fire – exposing their genitals as they do so – while the mistress of the house sits near by. But this depiction of snow is rare: in other early portrayals of winter, the landscape was never more than incidental to the labours themselves.

Bruegel modernized the tradition, chopping the number of frames from twelve to six, to reflect the number of seasons in the Netherlandish calendar, and starting his cycle in March. He also changed the focus of the works, so that instead of emphasizing the lives of the people and using the hills and fields as a backdrop, he emphasized their environment: the peasants are dwarfed by open skies, spectacular mountains and wide expanses of sea. This had tremendous implications. Viewers of these pictures would no longer see themselves as the most significant creation in God's universe; instead they were part of a larger and much more powerful world. Man's place had been usurped by the storm clouds of early spring, the blue of high summer, the gathering gloom of autumn and the grey of winter.

What made Bruegel turn his brush on the weather? It seems likely that it was the climate. The winter of 1564/5 had been the longest and most severe since the 1430s and the first great winter of the intensely cold period in Northern Europe that we now call the Little Ice Age. From 1520 to 1560, Europe had experienced relatively mild weather, but for the next 150 years the winters were exceptionally cold. This was the most sustained period of low temperatures in Europe since the last major ice age: crops failed, winter snowfall increased and Alpine glaciers advanced down the mountainsides, swallowing pastures, eradicating communities and gouging ever deeper features in the landscape. The people of the Chamonix valley, as the historian Emmanuel Le Roy Ladurie discovered, petitioned their lords to do something to alleviate the effects of the climate: 'We are terrified of the glaciers... which are moving forward all the time and have just buried two of our villages and destroyed a third.'

In England, the Thames froze over in 1565, as it would increasingly in years to come. The newly cold weather must have had a dramatic effect on an agrarian society that depended on the reliable arrival of sun and rain at certain times of year. The talk in the inns and from the pulpits and in the government would have been of the unusual weather. What better subject could the great artist take than a year-long look at the climate and its influence on people?

Of the five paintings of the seasons that survive – one has been lost or destroyed – Bruegel seems to have been particularly taken with the subject of *The Hunters in the Snow*. There are no known paintings of snow by Bruegel before 1565. After that, snow became one of his major themes. He painted at least four snow scenes in the following two years. *Winter Landscape with a Bird Trap* again depicts skaters – he renders their activities in much greater detail than in *The Hunters*. In *The Census at Bethlehem*, Joseph and Mary cross the town's snowy main street almost unnoticed. *The Massacre of the Innocents*, which hangs opposite *The Hunters* in Vienna, sets Herod's infanticide in a sixteenth-century village after a snowfall. In *The Adoration of the Magi*, Bruegel has moved the nativity to his own world and added thick blobs of falling snow which partially obscure the action. This is believed to be the first depiction of falling snowflakes in Western art, as well as the first nativity scene with snow in it.

None of Bruegel's later snow paintings was quite as brilliant and ground-breaking as *The Hunters in the Snow*. Some art historians have traced the tradition of landscape painting to this one image. Others have argued that the whole of abstract art begins in the white spaces of Bruegel's snowbank, in the empty blankness where there is nothing to focus on, no depth or perspective. If you stood this painting beside a Mondrian, one critic has written, the connection would be immediately apparent.

Bruegel died four years after painting his most famous snow scene. He had established an interest in the winter landscape that would spread across the Netherlands during the century and a half of cold weather that followed, and around the world thereafter.

Artists who became adept at the increasingly fashionable winter painting pioneered by Bruegel included Lucas van Valkenborch, Jacob and Abel Grimmer, Hendrik Avercamp, Esaias van de Velde, Isack van Ostade, Jan van de Cappelle and Aert van der Neer. As the Little Ice Age took hold and the winters became colder, snow gave way to ice, perhaps because there was less moisture in the air and so less snow. Frozen ponds and canals began to dominate. The strain imposed by the Little Ice Age may explain the bleak mood of the thirty winter landscapes of Jacob van Ruisdael. One shows a block of ice that has been cut out to make a hole for fishing: the ice is a full half-metre thick.

The number of winter landscapes produced in Holland declined in the second half of the seventeenth century, due partly to the economic depression, which obliged the wealthy art-owning classes to sell, flooding the market with works by older generations of artists. As the Dutch Golden Age drew to a close, the climate grew warmer again and remained so until the late eighteenth century. Rembrandt's *Winter Landscape* of 1664, one of the last of the genre in Holland, contains very little snow. But cold weather would return, and with it the artistic appetite for winter.

On leaving the Kunsthistorisches Museum I bought a postcard of *The Hunters in the Snow* and a poster as large as my grandparents' copy. We walked east through the great gates of the Hofburg Palace, then south towards Albertinaplatz, past bars and restaurants where people were sitting in the sunshine. We queued for an outside table at the Café Mozart and ate *Sacher-Torte* as we watched horse-drawn landaus drive past the Opera House. Then I pulled the postcard from my bag and looked at Bruegel's winter scene again. At that moment it occurred to me that there was something odd about the painting. Why, I wondered, was the ice exposed? The deck of cloud that fills the upper reaches of the painting is a dull bluish-green, and this colour is there in the ice on the ponds and rivers that run from the bottom of the hill to the shore of the distant sea. Given that a single track of animal prints is the only disturbance on the white surface in the foreground, the snow has clearly fallen recently, so surely the iced-over rivers and ponds would be covered with snow too? Bruegel must therefore be asking us to believe that it has snowed while the rivers and streams were still running, and then stopped long enough for the ice to grow so thick it can support the skaters. This seemed a highly unlikely sequence of events. Was it simply a mistake, then, and a major one, in this iconic depiction of winter? Bruegel would have painted it from memory in his studio. Had he forgotten that the blue-green of the ice would have been covered by the fresh snow?

I continued to wonder about the colour of the ice as we paid the waiter and set off east towards the Danube.

Where Bruegel and the Flemish and Dutch landscape artists who followed him had placed mankind within an influential but often benign winter, after the doctrine of the Sublime took hold, snow assumed a different character. When the cold weather returned to Europe in the late eighteenth century, it was no longer something for children to be seen playing with, but more likely to be shown freezing people to death, crushing them under its weight or drowning horse-drawn carriages in its hungry depths. In the growing romantic tradition in which nature was employed to dramatize and heighten human emotions, snow was to prove a useful tool, open to a range of sinister and dangerous interpretations.

The temperature in Europe began to dip after 1775, heading for a trough that bottomed out in the second decade of the nineteenth century. The winter of 1794/5 was harsh enough to allow a French cavalry officer to accept the surrender of the icebound Dutch fleet on horseback several hundred yards off the coast. It was snowing in Paris on 2 December 1804, the day Napoleon was crowned emperor of France, which was unusually early. In 1809, a series of major volcanic eruptions heralded the arrival of a particularly cold period as the clouds of ash partially blocked out the sun. The decade from 1810 to 1819 was the coldest in England since the seventeenth century. In 1812, a turning point in the Napoleonic Wars, the French Grande Armée was chased from Moscow by the advancing winter – known to the Russians as 'General Snow' – which claimed many of the 690,000 men who had set out on the invasion two months earlier.

In Europe and parts of North America, 1816 became known as 'the year without a summer', a consequence of the eruption of the volcano Tamboro in the East Indies. That year, crops and wine harvests failed across Europe. There was famine in Wales, and a frost every month of the year in Connecticut, which rarely freezes later than April.

The new coldness influenced literature and music as well as agriculture and politics. Even a cursory reading of Dickens's novels reveals that they are set against unusually harsh winters, which can be explained by the six white Christmases that young Charles experienced in the first nine years of his life (he was born in 1812). This may also account for the snowscapes vividly described in *Oliver Twist* and *A Christmas Carol*. Franz Schubert's song cycle *Winterreise*, or *Winter's Journey*, composed to the poetry of Wilhelm Müller in Vienna in 1827, is typical of the romantic interpretation of snow. *Winterreise* tells of a heartbroken man who walks miserably away from his life on a snowy winter's night. For Schubert and Müller, snow was the perfect metaphor for inner desolation, an exterior coldness to match the young man's desolate heart. *Winterreise*'s spurned hero has winter in his soul, which reflects the hard-heartedness of his lover: 'The girl spoke of love / Her mother even of marriage / Now the world is so gloomy / The road shrouded in snow.'

Art threw up two great romantic interpreters of cold. One was the German nationalist Caspar David Friedrich, born in 1774, who was given a taste of winter tragedy in childhood when his brother died while trying to save him from falling through the ice. Friedrich's beautiful, solemn wildernesses are full of religious symbolism: snow and death and transfiguration are closely connected. Beyond the fields of snow in his *Winter Landscape*, painted in 1811, redemption beckons a crippled boy in the form of a church looming out of the distant mist. But Friedrich's most powerful winter painting contains only a small amount of snow. In 1814, two years after Napoleon's catastrophic retreat from Moscow, Friedrich painted *The Chasseur in the Forest*, in which a French cavalryman walks, lost, into a wood of looming pine trees. A dusting of snow on the branches in the foreground foretells the soldier's death, and hints at the annihilation of the Grande Armée two years earlier.

The second great romantic interpreter of snow was the artist J. M. W. Turner, who, like Bruegel before him, travelled to the mountains of Europe to witness the full violence of snow and ice. He first visited the Alps in 1802, just a few months after his election to the Royal Academy at the age of twenty-six. On that and subsequent visits Turner saw the potential danger of snow at first hand. At least twice on his Alpine journeys his carriage was overturned by snow. He would go on to paint some of snow's most terrifying images.

Early in his career, Turner had sketched snow as a beautifying and soothing force, but as his art matured and his landscapes grew more tempestuous, his snow became more violent: it no longer lay meekly on the ground, but became an agent of power and fear. In 1810, he painted *The Fall of an Avalanche in the Grisons*, in which a chalet is obliterated by a mighty white wave descending from above. Two years later he produced *The Snow Storm: Hannibal and His Army Crossing the Alps* (1812), in which a vicious cloud reaches down from the clamouring sky to engulf the General's army, caught on an Alpine pass. Unlike earlier English landscape artists, Turner synthesized his works from several sources. In *The Snow Storm* he brought together a summer blizzard he had witnessed while staying with a friend in Yorkshire and mountainscapes sketched while travelling through the Alps. The result is a portrait of the Sublime – the most powerful of armies brought low by the forces of nature, just as Napoleon's was.

By the mid-nineteenth century the Little Ice Age was drawing to a close and, with the honourable exception of Gustave Courbet, who painted more than eighty-four snow scenes between 1856 and 1876, snow painting in Europe largely fell from favour. Towards the end of the century, however, the winter landscape made a comeback with a new group of celebrated interpreters to whom it was particularly suited: the Impressionists. The source of their inspiration was to be found half a world away, in a country that had remained isolated for two and a half centuries: Japan.

We meandered among Vienna's smaller streets, grazing on cake and coffee outside ornate cafés with uniformed waiters. Even though it was December, the blankets set aside for the use of customers were mostly left untouched, and the *Glühwein* at the winter market seemed an indulgence rather than a necessity.

At this time of the year, we couldn't walk a hundred yards without being offered a ticket to a concert by a man in a frock coat. Vienna seemed to exude classical music and high culture even on to its streets.

Back in London a few days later, I was trying to recapture that Viennese feeling when, on my way home from work, I stopped at a tiny back-street art gallery that specializes in Japanese prints. I had read about nineteenth-century Japanese art of the Edo period and seen the stylized, Tintin-like waves and snowscapes on exhibition posters. I was intrigued enough to go inside.

A dark-haired woman was on the phone, switching without any discernible effort between Japanese, English and German in the course of a single conversation. When she hung up I told her I was looking for winter landscapes, and wondered if she had any snow scenes. She laughed politely. I had come to the right place, she said. Then she showed me through some of the many boxes of original woodblock prints by the two masters of the form, Katsushika Hokusai and Ando Hiroshige, that were lined up by the window. She had an astonishing collection: there were more depictions of winter here than in any gallery I had visited. She had hundreds of original Edo-era snow prints, and many more digital reproductions. There were scenes of villages buried under snow, of snow falling on the broad straw hats and umbrellas of people hurrying through it, snow filling fields and resting on rooftops in soft, thick layers, and covering great mountains from top to bottom, as well as innumerable depictions of Mount Fuji's snow cone. Where Hokusai's palette was often reduced to three or four colours, Hiroshige's was rich and bright, with Prussian blue and red, and graduated areas where the blue faded to white. Each colour, the gallerist explained, was put on by a separate woodblock, on which the moistened sheet of paper was laid. Sometimes there were three, sometimes thirty or even fifty different blocks.

The colour of the snow came from an absence of ink, the plain white of the paper shining through. With a few dark lines on a blank sheet, Hokusai could create the impression of a village drenched in snow, almost erased by it. Hiroshige's snowscapes, meanwhile, demonstrated his mastery of light. He was able to capture, for instance, such nuances as the brightening of the sky after a heavy snowstorm has moved away. These specialists in the art of *ukiyo-e* emphasized the beauty of snow in a way that no Westerner had.

Why did they choose to depict snow so often? 'Snow is peaceful,' said the gallerist, resting a delicate finger on a patch of empty whiteness in one of Hiroshige's series, *100 Views of Mount Fuji*. 'Japanese culture values tranquillity highly. Not all of this type of art is so calm, there are some very noisy ones too, but this was an antidote to that, to the hectic business of life. They liked peace.'

Translated literally, *ukiyo-e* means 'of the floating world', which in Edo-era Japan meant geishas, sumo, kabuki theatre and all the indulgences of the newly affluent Japanese middle class. Asai Ryoi's novel *Tales of the Floating World*, published in 1661, explains his interpretation of *ukiyo-e*: 'Live for the moment, look at the moon, the cherry-blossom and maple leaves, love wine, women and poetry, encounter with humour the poverty that stares you in the face and don't be discouraged by it, let yourself be carried along on the river of life like a calabash that drifts downstream; that is what *ukiyo* means.' These beautiful portraits of snow, then, were a call to enjoy life while it lasted; to take pleasure from, among other things, the natural beauty of winter.

By the mid-nineteenth century, when Hiroshige and Hokusai were at their zenith, Japan had been cut off from the rest of the world for two hundred years under the military regime of the Tokugawa shogunate. When, in 1853, the American naval commodore Matthew Perry landed in Tokyo Bay and forced the government to open the country to trade, the nation's culture was released from its prison. Suddenly, the work of Japanese artists spread to the rest of the world.

In the 1860s, two exhibitions in London and Paris caused an explosion of interest in Japan. Western artists seized on the ideas and compositions that distinguished the woodblock prints, and in particular the work of Hiroshige, whom the essayist Mary McNeill Fenollosa later called 'the artist of mist, snow and rain'. Whistler vigorously promoted 'Japonism', as did Manet and Degas. Van Gogh collected woodblock prints, and Monet was so taken with *ukiyo-e* that he filled every wall of his house in Giverny with them and built himself a Japanese garden with an arched bridge inspired by a Hiroshige woodcut. Hiroshige's ability to play with light and conjure mood greatly influenced these artists, who came to be known as Impressionists.

The Impressionists' infatuation with the Japanese snow artists occurred just when there was a great deal of snow for them to paint. Between 1864 and 1893, northern France saw a series of winters during which snowstorms and subzero temperatures alternated with warm sunshine. The Impressionists hurried out into the frosty air with their easels and oils and created a genre of painting which they called *effet de neige*, or snow effect. Sisley and Pissarro produced numerous canvases in which snow or ice play a significant role, and Renoir, Gauguin, Caillebotte and Manet painted snow scenes too. But the greatest exponent of the genre was Monet. Of all the Impressionists, Monet painted winter the most, at the largest number of locations and in the widest range of light conditions, times of day and qualities of snow. Monet, a large man whose art it was said was only made possible by his exceptional constitution, took great pleasure sitting out in the snow, even in temperatures of −30c. When he accepted an invitation to Norway in the winter of 1895/6, he wrote to his friend Gustave Geoffroy, 'I painted today, a part of the day, in the snow, which falls endlessly. You would have laughed if you could have seen me completely white, with icicles hanging from my beard like stalactites.'

Monet's best-known *effet de neige* painting is *Magpie, Snow Effect, Outskirts of Honfleur* (1868–9). In it, we sit with the artist in a snow-covered field looking towards a rustic wooden gate, a stone wall and beyond it some trees and a farm building. He has captured a moment in time when a magpie has alighted on a gate in a snowscape transformed by afternoon sunlight. The surface of the newly fallen snow is heavy and warm, like whipped cream. We look towards the source of the yellowish light, which filters across the snow and through the trees, casting long shadows in the field where the snow is blue and purple and grey and occasionally white, and dark in the shadow of the magpie's small body. Here the Impressionist's palette, which in spring, summer and autumn contains a mix of bright colours, has been stripped down, and the result is simple and clear: the architecture of the composition, the geometry of a roof-line of a house, or a road, stands out. It is a remembered moment of silence and calm and peace.

The Impressionists did not look for symbols or hidden meanings. Their snow doesn't symbolize some aspect of human emotion. Their concern was not people's ideas about snow, but the snow itself. What they saw was something few artists had seen before: that snow is rarely, if ever, merely white. The angles and surfaces of snow crystals reflect and refract the different colours of sunlight that play upon them like the glass in a chandelier.

The Christmas lights were on in London as I continued my walk home from the Japanese gallery along streets that were bright with illuminations on a winter theme. There were snowflakes and snowmen in many of the store windows. Stuck to the glass frontage of a coffee shop was a sign written all in white, with a backdrop of stylized snowflakes. The legend read: 'Peace and Latte to All Men.'

What does snow mean to us now? We see it represented most often in sentimental, commercialized form, as kitsch. No one is invulnerable to kitsch, not even Monet, who captured the play of light on snow with such success that a painting like the *Magpie* appears on a thousand Christmas cards and biscuit tins, and has lost the power it once had. Even Dickens's winter, with snow poised so fetchingly on the panes of London's Regency windows, has become a cliché which has led to artificial snow being sprayed on acres of shop glass as a reminder of Christmases past. The snow globe and Santas with sleighs are quintessential kitsch, as is the great cardboard snow crystal hanging over book-shop tills. On television at Christmas, the tinkle of crystals falling to earth and the powdery flakes that fall between programmes are used to conjure an instant mood of nostalgia for our childhoods.

If commercial culture grew ever more infatuated with snow in the twen-tieth century, art became less interested, perhaps because the cold came to impinge less on us than it once had. Climate historians have clearly identified a decade from 1627 when Dutch artists simply stopped painting winter land-scapes. They believe that for a decade in the middle of some of the coldest years of the Little Ice Age, it became warmer, and for several years running there was not much snow or ice to paint. In the late twentieth century, snow art seems to have fallen similarly from fashion. There were exceptions, and perhaps there will be more as climate change once again has our attention.

In 2005, following a fact-finding voyage to the Arctic to examine the effect of climate change on snow, Turner Prize-winning artist Rachel Whiteread filled part of the largest space in London's Tate Modern with thousands of white plastic boxes. I walked among them, between shaped stacks of white which reached over my head, and thought of nothing so much as crumbling bergs. Whiteread had found a new context for an old meaning of snow, its transience. It was strange to think that this work was connected over a gap of 450 years with *The Hunters in the Snow* by its shared subject matter, in climatic circumstances that were diametrically opposite.

There is another modern-day reminder of snow in our art galleries. When we left the Kunsthistorisches Museum in Vienna, we walked back towards our hotel and stopped at the Leopold Museum, which contains several Egon Schiele paintings of Vienna in winter in the early twentieth century. Having spent so long looking at the Bruegels, I could not concentrate fully on these depictions of slum housing covered with dirty ice, and instead found myself staring at the white walls of the gallery, thinking about the white space where the hunters are about to tread, and about the white acres of minimalism. Snow is the colour of early-twenty-first-century living: we paint the insides of our houses white, as we do the walls of our shops and restaurants and galleries. The blank yards of white, with their absence of colour and depth, give us peacefulness, space in which to look, to think and reflect. We have brought the tranquillity of Hiroshige's and Hokusai's fields of snow into our homes.

In our house, I have hung my framed reproduction of *The Hunters in the Snow* on the wall near my desk. Sometimes I stand in front of it and imagine I can hear the children down there in the valley, playing on the ice.

It was on one of these occasions that I resolved the question of the blue-green ice that had baffled me in the Café Mozart. I recalled that during the hours I had spent in the Bruegel room a group of British visitors had entered with a guide. I had listened at a discreet distance as *The Hunters in the Snow* was introduced. An Englishman had said loudly that the mountains looked rather incongruous in what was supposed to be the Netherlands. The guide had quoted Bruegel's biographer, Karel van Mander, who wrote that Bruegel had 'swallowed all the mountains and rocks and spat them out again', after his return from the Alps, 'onto his canvases and panels'. The mountains were not incongruous, I had found myself telling the Englishman silently; it was just that he didn't understand the artist's decision to place them next to a Netherlands port.

I had focused on the painting in front of me, *The Return of the Herd*, Bruegel's depiction of autumn, in which the peasants drive the cattle down from the pastures for the start of the winter. The brightness of late summer was still visible on the left of the painting, but the dark clouds of winter were already coming into view on the right. Time, in other words, appeared to be marching across the canvas's sky from one side to another.

As I thought about this now, standing in front of my own *Hunters*, it suddenly dawned on me why the river ice was exposed – why it had to be.

I took in the painting once more, from the lovingly rendered still life of a blackberry bush to the Alps; from the town by the frozen sea far away across the broad plains to the watermill and the girls pulling sledges over the ice. As the Alps and the endless plains suggest, this painting does not represent a fleeting moment in a particular place, or indeed any view that has ever existed. What the presence of the mountains and the Netherlands coast in the same frame tell us is that Bruegel's landscape is universal. This is not a place; it is the world. Nor was there ever a moment such as this. This doesn't mean that skaters didn't play kolf, that people didn't gather on frozen ponds, or didn't extinguish chimney fires by throwing snow at them. But Bruegel has put different locations, times and activities into one document. He has made a movie of everything he wanted to portray about snow and winter and put it into a single frame. What he wanted to show about ice was the bare, blue-green colour that it sometimes reflected back at the winter sky, and that people skated on it. And how could he show skaters on a pond if it was covered with snow?

Instead of painting a real place at a real moment, Bruegel had shown a whole season. He had painted the entirety of that winter in 1565.

I went downstairs to tell Lucy. It all seemed obvious now.

CHAPTER 9

A Short Walk in Paradise

Rainier, Seattle, Glacier

A few weeks after our return from Vienna I was driving fast along the strip of flat land between the waters of the Pacific that lap the shore of Puget Sound and the Cascade mountains, blasting Beatles tracks too loud from the radio of the hire car. My destination was the ranger station at Paradise, on the southern flank of Mount Rainier, and high on coffee, jet lag and junk food, I felt myself to be on an optimistic, forward-looking coast where multi-trillion-dollar software companies were inventing the future. The buildings here seemed more modern, the smiles brighter and the people warmer than back in the east. Even the sun's rise over the land emphasized the sensation of being in a new and different world.

'Pencil, huh,' said the large man behind the counter at the gas station near Tacoma. 'Looks like a guy gonna do some writin'?'

'Sure am!' I said with my brightest, most caffeinated expression. 'Writing about the snowiest place in the world!'

'Huh!'

The urge to measure and calibrate snow, to compare one snowfall with another, had preoccupied me for years, but after encountering Bruegel's portrait of the coldest winter the sixteenth century had seen, I had developed an appetite for superlatives. I wanted to find the place that saw more snow than any other.

I had riffled through my books, dug through obscure reports on the internet and asked the weather experts I knew for leads. The world's snowiest place would not lie in the lower latitudes near the equator, of course, even if the mountains were high, because it would not be cold enough. Neither would it be too close to the poles, as they tend to be arid. The geographical requirements for heavy, regular snowfall put this place near a large supply of moisture, in the path of strong prevailing jet streams and a means of cooling the moist air within them: a coastal mountain range exposed to the world's prevailing westerlies would best meet these criteria. The ranges at the eastern edge of the Pacific, more than 60 degrees north or south, were likely candidates, which meant the southern end of the Andes, the Rockies in British Columbia, the Chugach Mountains in Alaska or the Cascades in Washington State. Having narrowed the search down this far, I discovered that Washington claimed to have two front-runners for the title, one of which was Paradise, the other Mount Baker. In the winter of 1972/3, the Paradise weather station recorded an immense 1,122 inches or 93½ feet of snow in one season, a record which wasn't beaten anywhere in the world for a quarter of a century.

What would it be like to live in such a place? How would people deal with so much snow? Where would they even put it all? Hungry for information, I searched online for photographs of the Cascades and found a website that broadcast live images from Snoqualmie Pass, at 3,000 feet. As I watched from eight time-zones away, a solitary car struggled uphill in the dark with its head-lamps blazing through grainy blobs that seemed to be big wet flakes of snow. Sitting in London that warm winter evening, I longed to be in that car, pushing up the hill in a storm, listening to the thrum of the snow chains as they found a grip on the soft surface.

So, early on a midwinter morning, I was steaming towards the snow again, and the volcano that dominates Washington State. It was further from Seattle than I expected: Mount Rainier's great size means that certain avenues in down-town Seattle appear to end at its feet, when in fact it is sixty miles away. As dawn turned to daylight, the mountain's cone turned from dusky blue to pink to white.

I drove through Ashford, past signs that read 'Volcano Evacuation Route' and 'Narrow Shoulder Next 15 Miles', 'Watch for Pedestrians, Bicycles and Elk' and various last stops for food, lodging, gas and tyre-chain rental, before entering the national park. I paid a uniformed ranger the entrance fee at the gates and began my ascent through the trees. There were patches of snow on the forest floor. Further on, by a sign warning 'Tire Chains Compulsory', the ground was white all over. As I climbed, the depth of the snow increased from a foot to a foot and a half, two feet, then three. Soon it was as high as the roof of the car, and as I crossed the bridge at the bottom of the Nisqually Glacier the snowbank was towering above the road, the temperature was plummeting and I could hear the wind singing around outside. The parking lot at Paradise lay deep below the surface of the snowfield, an acre of flat grey tarmac dug out from the surrounding ice crystals, and people with face masks, goggles, snow-shoes, snowboards and backpacks were arranging their equipment for a day in the snow. When I climbed out of the car, the cold seeped easily through my thermals and jacket. I put on my snow boots, hat, hood and gloves, and set off up the hill.

After a short distance I found the old Paradise Inn, a tall stone-and-timber structure built in the alpine style in 1916. It was almost completely buried in snow. When Floyd Schmoe, one of the first rangers to be posted to Paradise, arrived here in 1921, exhausted after a difficult climb through the drifts from the foot of the mountain, he'd had to tunnel down to an upstairs window to get into the building. Schmoe had been sent from Seattle because the National Park Service worried that the two caretakers they had left there in the autumn to keep the inn from being crushed by snow had tried to kill each other during their enforced cohabitation. One of the men had stumbled and swum through the snowdrifts to arrive, confused and battered, at the ranger station at the mountain's foot. Schmoe found no sign of the other employee at Paradise. Had he died in the fight, or in the snow? Or had he deserted? Schmoe did not waste much time worrying about him: he had more urgent things on his mind that winter. When he heard the beams of the inn creak, he must have imagined the weight of snow bearing down on them and feared that he would be buried in the place he had been sent to safeguard.

When I arrived, the rangers had cleared the way to an oval steel doorway which was sealed with a heavy padlock. The inn was being refurbished, so I took a walk around the outside. The snow had a thick crust, which broke in slabs under my boots. It felt strange to be at roof-height, suspended by the deep white ocean of crystals beneath me. I remembered the words of the snowboarding pioneer Tom Sims, who said that riding the snow in the mountains of Washington was like surfing the Pacific; it was even the same water. I was circumnavigating this large hotel while it was submerged in the recycled waves of the nearby ocean.

When I had completed the lap I climbed back into the snow trench that marked the track to the car park, and walked on to the modern visitor centre, which was also buried. A willowy young man named Naaman radiated rangerly sanctity from behind the reception desk. Naaman had been at Paradise for a couple of years, he said. From the back room he produced a red binder full of snow-depth data stretching back to 1948. In 1972, they'd had almost thirty feet of snow on the ground for most of the winter. Last year had seen the fifth-lowest snowfall in recorded history, but this year so far they'd had 538 inches – way above the 443-inch average – since 1 July, so they were doing well. 'We can get totally buried here,' Naaman said. He disappeared into the back again and returned with some photographs that showed the Paradise Inn in various states of submersion through the decades. If I was interested, he could take me on a snowshoe tour of the area with some other visitors. An hour later, I was strapping on some old-fashioned, tennis-racket snowshoes and following him west towards the Nisqually Glacier.

It was incredible that such a tiny thing as a snow crystal could have so powerful an effect on the environment, Naaman said as he led us across the mountain. But snow affected everything: the animals, the plants, even the land-scape. Because Rainier received so much snow, there were twenty-five glaciers flowing down its flanks. The combined volume of ice on the volcano had been estimated at a cubic kilometre, and each glacier was sculpting great valleys and troughs, grinding and bulldozing its way downwards. After some time we reached the edge of the canyon carved out by the Nisqually Glacier and stood on a precipice looking up to the summit of Mount Rainier 9,000 feet above. The glacier, which ended in a small stream of meltwater in front of us, used to extend right down to the road bridge a thousand feet below, but it was receding, as was all Rainier's ice. Even so, the mountain's glaciers were the main water reservoirs for the millions of people who lived in the conurbations of Seattle and Tacoma, Naaman said. The snow stores water that would otherwise run straight off the land into the sea.

As we flapped our ungainly way on over the crunchy white crystals Naaman spoke about some of the other ways in which life here depends on snow. Snow protects the flora and fauna from the extremely low night-time temperatures that descend on the mountain in winter, acting as a winter coat. The animals and plants that live up here have all developed different ways of living with snow. The Rainier marmots, for instance, have become much more sociable than other, lower-altitude breeds because they need to be: in winter, they hibernate together for warmth and will be in danger of freezing if the snow isn't deep enough to provide insulation above their burrows. Their pulses slow from 60bpm to around one beat every ten seconds. They also have a highly developed spatial memory: they can wake from hibernation, walk several feet over the snowpack and know immediately where to dig to find the entrance to one of their tunnels.

Pikas are a type of rabbit that has adapted to the cold by losing its big, floppy, blood-cooling ears and developing tiny ones in their place; during the long Rainier winters, they live off the bundles of hay they collect in the summer. Sometimes, for warmth, they sleep next to the marmots, who are in such deep hibernation that they are oblivious to them.

Mice and stoats also dig tunnels through the snow. When the snowpack is light it consists mostly of air, but when they find it more compacted the animals scurry back up to the surface. Foxes, meanwhile, have extremely bushy tails which serve as blankets. There are also elk and mountain goats on Rainier, as well as black bear which sleep for most of the winter, waking up once in a while to forage for food. Because their feet are not broad enough to allow them to walk in heavy snow, the deer migrate down to the valley when the snow comes, with the cougars in pursuit.

Plants, too, show remarkable adaptations to the harsh mountain environment. The trees that grow at high altitude on Rainier are specialists at dealing with the low temperatures and blowing ice at Paradise. The two main varieties, subalpine fir and mountain hemlock, adapt to the snow in very different ways: the hemlocks are shorter, with long, floppy branches which bend under the load of snow so that it eventually falls off and the branches spring back up. The firs, meanwhile, have stiff branches that grow close to the tree's trunk, and its tall, slender profile means that it sheds much of the snow. Because of the short growing season, the trees reproduce asexually, so are found in clumps, with the grandfather tree in the centre sheltered from the wind by the second- and third-generation trees growing around it.

The snow also protects from frost the flowers that grow on the meadows in the short summer months: some survive dormant for years beneath the snow before, one hot summer, it melts and they seize their moment to bloom and reproduce. The avalanche lily in particular is equipped with growth tips which are warmed by the plant's own metabolism and melt their way up to the surface. When at last it breaks through, the lily unfurls white voluptuous skirts and yellow legs above the snow. The snowpack is home to a great number of invertebrates, too: ice worms, grylloblattids, daddy longlegs and snow fleas have all developed the ability to live there on the bounty of insects which are carried by the wind on to the mountain every day.

Snow affects everything up here, from the trees to the hunting patterns of the cougar, even the shape of the mountain. 'Everything is connected,' Naaman said.

After Naaman's talk, I made my own way back through the trees to the ranger station. At the top of a low rise I looked out over the dark foliage of the hemlock and the firs, draped with a sprinkling of heavy snow. The mountain fell away into a deep valley, before rising again in a serrated ridge, the remains of an older volcano that had been thrown up by the heat of the Earth, then worn down by the action of snow and ice, just as Rainier itself would be, with its many frozen rivers grinding invisibly downwards.

❄

In the early summer of 1999 a group of experts in extreme weather drove east along the dead-end Mount Baker Highway, a hundred miles north of Paradise. The highway wends its way inland from the Pacific between vast Douglas firs and ancient Western cedar trees, past hippie homesteads and farms, into the Cascade mountains. The road steepened as it climbed towards the ski area on the eastern flank of Mount Baker, the height of the snowbank growing rapidly as the team gained altitude. In the forest to either side of the road, they noticed that giant trees, which had stood for well over a century, were lying at strange angles, some pushed over completely by the creeping snowpack. When they arrived in the car park at 4,200 feet, there was almost twenty feet of snow on the ground. It was the second week of June.

According to Mount Baker's weather station, this was the end of the snowiest winter season ever recorded. More than 95 feet of snow had fallen since the end of July the previous year – enough, if it hadn't settled or melted, to bury a ten-storey building. This was 18 inches more than even Paradise's record snowfall of 1,122 inches in 1972–3. The team from the National Climate Extremes Committee (NCEC) were followed by news reporters and TV crews: if the committee verified the record, this was an event of international significance.

As well as resort personnel, the NCEC experts interviewed the observers who measured the snowfall. They also interviewed the plough drivers from the Department of Transportation and examined the measuring equipment, reviewed the snowfall records and inspected tree damage caused both by avalanches and by the weight of the snow as it crept downhill. They then left

Mount Baker to write their report, which they submitted to the director of the National Climate Data Center (NCDC), Tom Karl. The rangers on Mount Rainier, along with the media, waited on tenterhooks to hear the NCEC's result.

It was not a straightforward decision. The NCEC report reveals a number of anomalies in the methodology of the weather station operators at Mount Baker, such as the fact that the snow depths were not always measured in the same place. These were 'generally' taken in the parking lot next to the employees' building, a site which was far from ideal because, the report states, 'after the snowpack accumulates to several feet or more, the plowed parking lot increasingly begins to act as a depression or pit, which this year [1998/99] reached a depth of 20–25 feet'. The effect this would have on the snowfall measurements would be significant. Measurements in the lot were typically taken on the asphalt, but sometimes on a snowboard, 'with occasionally utilization of the hoods of vehicles'. The committee's most serious doubts appear to have sprung from the fact that the records of the Washington Department of Transportation, whose plough drivers took regular measurements adjacent to the ski area, were 10 to 29 per cent lower than those of the ski personnel in all of the six months in which comparisons were made. In a section titled 'Areas of Concern', the report states: 'Many ski areas have low credibility for accurate measurements since economic self-interest is served by reporting favourable snow conditions.' Could it have been, then, that the ski resort was inflating the figures artificially in order to claim the record?

After leaving Paradise I tracked down Robert Leffler, the man who had led the NCEC team that had been sent to Mount Baker, at his office in North

Carolina. I asked him why they had been sent in the first place. Had the NCEC been suspicious of the resort's claims? Not at all, he told me. Most of their data came from sites supervised by the National Weather Service and, since this was not such a station, they simply needed to verify the procedures that had been used. Even the NWS-affiliated stations were monitored. 'Quite a few ski resorts have supervised stations, but we have to keep a close watch on some of those too, because there are concerns about conflict of interest.'

In the end, the team decided that the ski resort's claim was authentic. 'We had several cross-checks,' said Leffler. 'For instance, in the old days there was a published NWS station at Mount Baker, and the data from that showed that the mountain did sometimes get higher snowfall than Paradise. Plus we could see the amount of damage the snow had caused to the forest – some 100-year-old trees had been knocked over by snow creep, which showed this was a very unusual snow event. And when we visited the site the amount of snow still on the ground in June was another indication that made us feel comfortable about the decision.'

On 2 August 1999 Karl accepted the team's recommendation and verified the new record. Baker had officially defeated its arch-rival, Paradise.

Baker celebrated, but the rangers at Paradise, who had been dutifully logging the snowfall in the manner prescribed by the Weather Service for the better part of a century, were frustrated: they had just seen their title as the world's snowiest place usurped in a local derby. 'There was some reluctance that the committee should accept a record from a non-Weather Service affiliated site,' Leffler said. 'They were bound to feel that. Wouldn't you?'

Naaman, when I asked him, made light of losing the snowfall record to
Baker. 'We did think about cloud-seeding,' he said, through teeth that might
have been slightly clenched. 'But in the end there wasn't much we could really
do about it.'

'Buyin' a map!' said the sun-wizened woman at the gas station on the outskirts
of Seattle the following day. 'Must be goin' travellin', huh!'

'Sure am!' I beamed. 'Heading to the snowiest place in the world!'

'Huh!'

The map told me that to reach Mount Baker I should drive north towards
Bellingham and turn east ten miles short of the Canadian border. First, though,
I had an appointment with the National Weather Service in Seattle to try to find
out why Baker received so much snow – and why the winter of 1998/9 in particular had been so extraordinary.

The NWS office overlooks Lake Washington on the eastern side of Puget
Sound. Through the vast windows I caught sight of a postcard view in primary
colours of the lake, the grass and forested knoll on the other side, and in the
distance beyond the snow-capped peaks of the Cascades. It was a cold day:
winds from the north-east had brought the temperature down, but the sky
remained unsmudged by a hint of cloud right across the state. I knew that
because Jay, the quick-talking senior meteorologist, showed me the live satellite image on the bank of screens in front of him. It was extremely dry at the
moment, Jay told me, unusual for the time of year. If I was looking for new

snowfall I was going to be disappointed. It was so dry that it reminded Jay of his previous posting on the desert plains of Nebraska.

He clicked the mouse and a window on the screen popped up. This showed the Doppler radar, which did a 360-degree sweep every ten minutes or so, and each time it did a scan it came up with wind speed and direction at different altitudes, so he could see if there was a storm coming. Right now, there wasn't. He pulled up another window, in which were a series of time-lapse photographs showing the whole state, taken during the past twenty-four hours from a satellite 23,000 miles up. The snow in the mountains stood out clearly. He ran the sequence, and the different angles of the sun throughout the day made the snow appear to wriggle on the high ground.

The secret of Baker's snow lay in the way moisture arrived at the mountain from the Pacific. Moist westerly airstreams funnelled around the 5,000-foot mountains of Vancouver Island, and these streams bowled down the Strait of Georgia and the Strait of Juan de Fuca until they met at the island's south-east corner. There they merged into a single powerful stream which rushed eastwards into the Cascades. The mountains pushed the air upwards, cooling it and forcing moisture to condense into clouds, which then dropped immense quantities of snow. The Cascades at this latitude are a hundred miles deep and high enough to slow or even stop the airstream. They were so efficient at taking moisture out of the air that by the time it had crossed the mountains there was no moisture left: Spokane, which lies east of the high ground, is in a desert.

What was it about 1998/9 that made it such a special season? It was most likely down to El Niño, Jay said. During an El Niño year, warm water travels east across the Pacific to the coast of South America where it is forced north up the California coast. Meteorologists had observed that El Niño had the effect of splitting the jet streams of moist air coming in to the West Coast of America: one stream would tend to head up towards Alaska and another down to California, leaving the Pacific Northwest with only a weak system. However, during the opposite effect to El Niño, La Niña, when the water in the central north Pacific is cooler than usual, Washington tended to get the full jet stream, which just kept on going and going. The year before Baker's record-breaking season saw one of the strongest El Niños ever recorded, with warm water flowing from South America all the way up to Washington. 'In the summer of 1998, before the record snowfall year in Baker, the temperature of the sea in the North Pacific was absurdly high,' Jay said. 'They were catching tropical fish off the coast here.' Then a strong La Niña kicked in, and the ocean temperature swung rapidly the other way. 'We transited into La Niña pattern very rapidly that winter.' It was just conjecture at this point, but Jay suspected that the warmer sea surfaces caused by the strong El Niño that summer had released an unusually high level of moisture into the atmosphere over the Pacific. The swift transition into La Niña had driven this on to Washington's mountains, and Baker in particular.

The year of 1998–9 produced record snow for much of the Pacific Northwest, not just for Baker. Automated data recorders built into the mountains reported 200-year maximum snowfalls in several places across the region. Paradise received its second-highest snowfall ever, at 1,035 inches, just 105 inches fewer than Baker.

❄

From Jay's office I followed the interstate north along the coastal plain, under flyovers that were heavy with vegetation, then turned east on the Baker Highway. At Glacier, the last inhabited village, which consists of ninety souls, a store, a couple of bars and some ski-hire shops, I stopped to buy food and then pressed on the few miles to Dee's house, which stood on the flank of a low hill between the enormous Douglas fir and cedar trees. Some of the cedars up here can live for more than a thousand years.

Dee, who was in her sixties, was an inveterate snow person. As we talked about snow over the course of the evening I fell a little bit in love with the life she led out in the woods, beneath the mountains.

She and her husband Steve were keen skiers who used to come up from Seattle for weekends, and when Steve retired they decided to move here for good. They had cleared a plot of land of trees and the huge boulders that had fallen off a nearby mountain, and built themselves an elegant modern home of wood and stone. Steve, who was working on a project in Seattle that night, was a fine skier and had been in the patrol at another ski area. He liked to go into the back-country, and Dee put on a DVD with photographs of them and their friends climbing up ridges and coming down through the fresh powder which reached up to their chests. Sometimes it looked as though they were almost drowning in it. The trees in these photographs were so heavily caked with snow that their branches bent down to touch the snowpack.

It was a shame, she said, that I had come at such a bad time for the snow. There hadn't been any for at least a week.

That evening, we drank tea and watched winter sports for a while on a vast screen high on the wall. Then Dee became my travel agent. She contacted a friend who worked for the company that owned the ski area and organized a guided tour for me first thing the next day. In the morning I should go and see John, who had a ski-hire shop, to borrow a board and some boots.

We talked about snow long after night had fallen in the forest outside. I went to bed with the scent of pines and frozen water in my nostrils, and dreamt of stepping out in bright sunshine across the boundary rope, on to the virgin mountain.

The next morning I hired a board in Glacier and headed east up the highway towards the ski area. Some miles up the road, white flakes began falling on the windscreen out of the clear blue atmosphere. I stopped and climbed out of the car, and tried to catch one in my hand, but it fluttered away, so I picked one off the bonnet to examine it. It was not a snowflake but a white seed-pod, drifting down from the great Douglas fir. The ancient forest was regenerating itself.

At the White Salmon Day Lodge I met Andy, the director of weekday ski patrol, a tall, taciturn man who took his job extremely seriously. He had left Oregon for college in Washington a decade previously and had since worked every winter season in the ski area. He was evidently drawn to the empty country up here, and initially when he spoke the words came out grudgingly, as though the talking interrupted other, better things that were going on in his head.

We took a series of chairlifts 7,000 feet up the side of the mountain to
the hut at Panorama Dome which faces the summit of Mount Shuksan. Near the
top of this 9,127-foot mountain a hanging glacier seemed to be on the brink
of a vertiginous fall. Inside the hut, which was built on stilts, two young ski
patrollers sat in a silence broken occasionally by the crackle of walkie-talkies,
their feet up, eyes closed behind Ray-Bans to deflect the glare that poured
in through an immense picture window. There hadn't been a snowfall for
several days, and as it was midweek the resort was quiet. On the hut's rough
wooden walls was the paraphernalia of the ski patrol's trade: a rucksack, greasy
tools for fixing mechanical lifts, ropes and carabiners, as well as a tin of smoked
oysters.

'Remember that time a few weeks back when we were just getting our
butts kicked?' Andy asked the others. It had been before the Superbowl and the
Banked Slalom snowboard race. The mountain had been hit by a major storm
system that had dropped more than a foot of snow each day for a week: in
total 110 inches, or just over nine feet. On days like that, people skipped out of
their jobs in the cities on the coast and hurried east along the highway in their
pick-ups and four-by-fours, hoping to make it up the mountain.

During Baker's record-holding season it had snowed almost non-stop for
the months of January and February. These were lucrative times for the resort,
but they were also the hardest times for the patrol. During snowfalls like that,
they started work in the dark and finished in the dark. Snow buried everything.
It barricaded the entrances to the lodges, which had to be cleared, and clung to
the lift towers, which had to be shovelled off. Everything that was left outside

had to be dug out: snow-machines, cars, snowcats. Even the platforms people stood on to get on the chairlifts, built thirty feet above ground level to allow for the depth of snow, were submerged. It was hard sometimes to find places to put all the snow, which would lie until the spring: some of it was pushed out on to the surface of a frozen lake, and some into deep hollows in the mountainside. Over the season these filled up, so that where there had been depressions on the mountain there were now bulges. For two days in the middle of February, the snowfall had been so heavy that the ski area had to be closed for the first time in its thirty-three-year history.

Keeping the access road open was an enormous task too. Driving up to Baker during the winter of 1998/9 was said to be like driving through a steep, never-ending tunnel with only a strip of light at the top between the immense snow cliffs. The snow on either side of the road had built up so high that the snow-blowers owned by the Washington Department of Transportation couldn't fire over it, and special equipment had to be rented to lower the snow-banks before a path could be cleared between them.

But the patrol's biggest job was keeping the customers safe. This meant touring around the edges of the resort digging out the boundary ropes so that people knew where the limit of the pisted area was, and trying to prevent people falling into tree wells, the areas of soft snow beneath the branches of the trees. The snow here can turn into a quicksand that skiers can drown in.

At Heather Meadows, Andy let us into the old lodge, which was only open at weekends, taking us in the back way, through a janitor's cupboard and the men's toilet. Inside, the walls were lined with photographs recording Baker's history. There was a 1927 brochure for Mount Baker Lodge, 'the newest national playground', just '60 miles from Bellingham, Washington', with bungalow cabin rooms from $5.50 per day. There was a photograph of Clark Gable, Loretta Young and Jack Oakie in the 1935 film adaptation of Jack London's *The Call of the Wild*, which was shot down the mountain on the Nooksack River. Then there were historical pictures of the snow, which Andy could date by the models of cars in the parking lot. One showed a snowbank at least twenty-five feet high overhanging the lot, with 1950s cars parked beneath it, and another, from the late 1960s, under a snowbank of similar size. There were photographs from the record-breaking winter of 1998/9, too, and a photograph of the National Climate Extremes Committee verifying the snowfall.

Outside the lodge was the snow stake, the point where the record snowfall had been officially certified, and a display board, showing how much snow had fallen in the past twenty-four hours, which was swept once a day. Andy measured it every morning with a ruler. There was also an electric eye positioned above it whose readings were sent direct to the Northwestern Weather and Avalanche Centre at the National Oceanic and Atmospheric Administration (NOAA) offices in Seattle. The total depth of snow was measured by another electric eye, above our heads, and manually by a black-and-white snow stake, next to it.

The snow on the pole now stood an inch below fourteen feet. In 1998–9 they'd had to add another ten feet of pole to the top of this one, which already reached way above our heads.

In the end, of course, it was a near certainty that neither Baker nor Paradise was the snowiest place in the world: establishing such a thing would be impossible. Measuring snow, with its tendency to settle and to move with the wind, to vary the speed at which it melts and metamorphoses depending on the surface it lands on and the angle at which the sun hits it, requires a set of complicated rules and procedures. These must be applied to very different geographical features and adhered to throughout a whole snow year. There is neither the money nor yet the public interest to justify placing and maintaining sensitive measuring equipment on the remote mountaintops where it snows most. There is no weather station at the summit of Mount Rainier or Mount Baker, never mind in the remotest parts of British Columbia or the Chilean Andes. Even the weather station at Thompson Pass outside Valdez was closed.

I thought again about Matt. He was convinced that Thompson Pass got more snow than anywhere else, and after my warm week in the Pacific Northwest I began to think he might be right. Even though the weather station there had been shut down three decades ago, the pass still held seven of the ten snowfall records listed in the NCDC extreme events. Matt, who had a snow stake in his garden and measured the snowfall and temperature daily, said he thought that certain of the glaciers at 5,000 feet got more than a thousand inches a year on average, which if true would put them above both Baker and Paradise weather stations. But without a certified measure, no one really knew.

Bob Leffler referred twice in our conversation to Mount Rainier as being 'the snowiest place in the world'. Hadn't he been to Washington precisely to prove that Baker was the snowiest place? Yes, he'd verified the record for one snow season, but on average, over the period for which figures exist, Paradise received more snow than Baker. And it was likely that Paradise would win back the annual seasonal snowfall title before too long, he said.

Snow, it seemed, was going to elude the grasp of those who attempted to pin it down for some years yet. But for now this patch of level, snow-covered ground behind a ski lodge would have to do; it was here where, barring the occasional reading on a car, they had recorded the highest winter snowfall in the world.

I took a photograph of Andy at the snow stake, then we shook hands and parted. I was eager to do some snowboarding and caught the lift back up the mountain. It was almost a year since I had been out on a snowboard; it had been in the Alps, after leaving Philippe.

As I contemplated the near-deserted slope before me, laid out beneath the glittering summit of Mount Shuksan, I thought back to other superlative snow I had experienced. There was a day in the Grand Tetons in Wyoming, when we had skiied from the top of the mountain to near the bottom at breathless speed and taken a turn into the narrow confines of a frozen creek. I had found myself suddenly alone and stopped, sweating and out of breath, to wait for the others. Big flakes the size of silver dollars tumbled through a silence broken only by the whisper of the landing crystals, which brushed my cheeks and caught in my eyelashes, and I thought how lucky I was to be there. And there was the time, one morning in Chamonix after a heavy snowfall, when we had caught the cable car from Argentière. As we climbed down the steel steps from the highest station we could see north and west for fifty miles. We had descended through snow so deep we were surfing, and when we fell over it was almost impossible to get up again because of the sugary softness of everything. I could remember the hiss under my board as I turned slow S-shapes on the sea of new crystals.

At Baker, I set out cautiously, unused to the length of the rented board or the heaviness underfoot. It took several runs before the heat of exertion began to spread through my body. On my fourth descent I cut off into some fresh snow among the trees and slalomed down between them, flashing through the long black stripes of shadows, then out across the wide flat run-out as fast as I could go. When I reached the end of the piste my eyes were streaming. I looked back at the track I had made in the snow, and at the mountain and the trees and the blue sky above, then hurried over to join the queue to the lift that would take me back to the top.

CHAPTER 10

The Last Snow in Scotland

Aviemore, Cairn Gorm, Garbh Choire Mor

The weekend after I returned from Washington was so unusually cold that it snowed in London. The flakes fell thick and deep enough to last for most of Saturday, when Harry, Arthur and I went out into the park and made a pair of snowmen. We were not alone. By mid-afternoon, the expanse of flat ground was filled with an army of standing snow creatures, equipped with gloves and hats and twigs. It had been an exceptionally cold winter in many parts of the world. It had snowed in Baghdad, there were great blizzards in China, and the Arctic sea ice had been widespread. Some analysts put this hiatus in the warming trend down to La Niña, others to sunspots, or simply the normal variance of the weather.

There had even been an abundance of snow in Scotland, which had seen its best ski season for at least a decade. The Highlands, where skiing had begun for me, and from where I had developed a strong enough connection with snow that I had once carried some home in a thermos and put it in the freezer as a souvenir, had been suffering from a dearth of the stuff for some fifteen years. When I had last visited the Cairngorms, one January in the 1990s, there had been barely any snow at all. A few ardent skiers made the journey up the hill regardless, but where there should have been pistes, streams ran open and the rocks and heather were exposed to a lashing wind. That was the year I abandoned Scottish skiing, as almost everyone else had done, for the more reliable slopes of the Alps. Now, after seeing the extreme snow of Paradise and Mount Baker, I wanted to go back to find out what had happened since I had left. It was almost the end of the snow season, and this would be my last journey.

I caught the Caledonian sleeper from Euston. The train must have once been a glorious addition to the rolling stock of the old British Rail, but now it had become shabby. The carriages and staff shared a careworn appearance, as if they had started out brightly together in the 1970s and had been plying the tracks sleeplessly from London to Inverness ever since, repaired but never refurbished. Nevertheless, it was a romantic way to travel, even alone, dozing in a bunk while the rump of Britain passed slowly by in the darkness beyond the compartment window. As I drifted in and out of wakefulness, I felt the journey's slow rhythm, the train halting at some unknown northern junction before starting forward again with a squeak of springs and an accelerating clack-clack. Was the driver chosen for the special care he took? Perhaps the measured pace of his movements was simply habit, the result of driving this sleeping cargo through the darkness, night after night, year after year, for decades.

Shortly after dawn, a steward knocked at the door with breakfast in a white cardboard box. We would be arriving in half an hour, he said. When I opened the blind the train was travelling through heather and I could make out cars overtaking us on the A9 near by. Through Glen Garry and Glen Truim, over the Drumochter Pass, the road and railway track run close together, rarely more than a few hundred feet apart. Past Dalwhinnie, as we entered the Spey valley, the horizon drew away rapidly to the east and soon Cairn Gorm was visible several miles off, with a covering of snow that reached down to its midriff.

When I first caught sight of Britain's largest mountain range as a child I found it underwhelming. I had expected rocky peaks, but looking across the valley from the other side of the Spey, the Cairngorms appeared to be little more than a giant moor with the sort of rounded top I had seen in the Yorkshire Dales. There was none of the drama of the Lake District crags. People who lived in the Highlands called them hills too, even though five of the six highest summits in the British Isles are among them. It was only when I was delivered halfway up Cairn Gorm in a bus that I realized how big they were and how quickly bad weather could descend on them.

I left the sleeper at Aviemore and carried my bags along the platform and out on to Grampian Road. The village had changed in subtle ways since my last visit. Pizzaland, where we had once loaded up on cheap carbohydrates and dared each other to eat forkloads of hot chillies, was gone. The garage had been closed and acquired for development, and several buildings at the southern end of town had been demolished. Most surprisingly, there was a security guard on the long entrance ramp up to where the Aviemore Centre had been. Were they now trying to keep people out? I walked that way out of curiosity, past a bright new sign with an unfamiliar name: Macdonald Aviemore Highland Resort.

The Aviemore Centre was once the crown jewel of the Scottish ski industry. It was built during the sixties ski boom by the department-store tycoon Hugh Fraser in an attempt to grab a share of the spending power of the quarter of a million or so Britons who were beginning to flock to France, Switzerland and Austria. Fraser's ambition was to turn Aviemore into the St Moritz of the Highlands. The master plan incorporated clothing and sports shops to equip the new British skiers, hotels, restaurants and a cinema to house, feed and entertain them, and a giant ice rink and dry-ski slope for the days when the weather closed the mountain. There was even an electric revolving door, the first one to reach the Highlands, at the entrance to Cairdsport ski shop.

Fraser didn't live to see his grand project opened: he died in November 1966, a month before the ceremony, which was then conducted by his widow. But he should have been pleased with his creation. Three-quarters of a million people visited the centre in its first year, many of them learning to ski for the first time on Cairn Gorm. In 1968, the Earl of Dundee declared in the House of Lords that 'There is no doubt that this is one of the best skiing areas in Europe.' In the seventies and early eighties, A-list entertainers performed at the concert hall and members of the royal family visited from nearby Balmoral. It wasn't until the end of the eighties that decline began. By the early nineties it looked terminal. The concrete slabs that had been used to pave the resort's public spaces and walkways had begun to break up and sprout weeds. The artificial ski slope's steel chassis had rusted and curled up at the edges, threatening to carve up anyone who dared take it on. The hotels became dilapidated, and the bars and clubs that had spilled out rowdy crowds at Hogmanay were closed or empty.

I walked around the new Macdonald Highland Resort trying to recall where the old landmarks were. Where the ice rink must have been there was now a car park. The village of chalets, one of which we had rented for a week in the seventies, was gone too. The dry-ski slope had been removed, and even the hill on which it had run seemed to be lower than I remembered. Near the site of Cairdsport was a new mall called The Brands Are Gathering, a large self-service restaurant and a conference centre. The new complex was smarter, more corporate. It was evidently chasing a different clientele, one whose needs were better served by flip charts and whiteboards than by skis and ice skates.

Local people are said to have cheered when parts of the Aviemore Centre were demolished in 1998. It had reached such a state that it had been dragging down the reputation of the village, they said. To me, though, the destruction of the centre marked the end of a time when there had been a hope, even an ugly, concrete one, that Britain could sustain an indigenous skiing industry. There were several good reasons why Fraser's dream of a Highlands St Moritz had been ground out of the country, among them the growth of cheap air travel and the development of the Alpine resorts. But chief among them, it seemed to me, was the lack of snow. It could not be a coincidence that the year the centre was demolished was the warmest on record.

At one time, the area had been so popular with skiers that the not-for-profit Cairngorm Chairlift Company couldn't think what to do with all the money it was making, so they ploughed it back into the lift system. Even after the run of mild winters in the late eighties the company was keen to expand the business, and it was only the planning authorities that prevented them from building more lifts and access roads. This bureaucratic brick wall may have saved them in the end: other Scottish resorts that had spent heavily in the nineties went bankrupt in 2003. Even in the nineties there had been 150,000 skier days per year on Cairn Gorm. By 2006, all of Scotland's five ski areas together could barely boast that many. In 2007, Cairn Gorm saw just 38,000 skier days.

Scotland's snow is not uniquely fragile. A recent study of snowfall in the Swiss Alps showed a dramatic drop since the late 1980s, with snowfall in some years 60 per cent lower than in the early 1980s. Monthly snow-cover in the northern hemisphere, where the great majority of the Earth's snow lies, has decreased by 1.3 per cent per decade on average for the past forty years, and climate models predict a reduction of 60 to 80 per cent in snow overall by the end of this century. Ironically, more snow may now be falling on high ground as precipitation increases because of the extra moisture in the atmosphere, but snow-cover as a whole is expected to decline greatly, even in areas of vast snowfall such as the Pacific Northwest. An increase in midwinter temperatures of 2C would reduce snow-cover in the Cascade mountains by more than 20 per cent.

But the Scottish mountains' low altitude and relative warmth, sustained by the effect of the Gulf Stream, make the snow that falls here particularly vulnerable. In the forty years leading up to the winter of 2004/5, the average number of days of snow-cover in Scotland declined by a third. By the 2080s, meteorologists predict that Scottish snowfall could be a tenth of what it was at the start of the twenty-first century. Faced with these facts, the Scottish executive made plans to offset the demise of winter sports by encouraging other leisure pursuits that are not dependent on snow.

As I travelled around the Aviemore area I could see that diversification was already well under way. Many new businesses had sprung up which were not connected with snow. The village of Newtonmore had gained a water spectacular called Waltzing Waters. There were at least two new mountain biking centres. The five-star Hilton Coylumbridge Hotel, designed as a swanky base for skiers in 1965 by the Rank Organisation, had built a huge indoor children's play shed. Further up the road, the main attraction of the UK Sledging Centre, which sat just below Britain's largest ski area, was its dry-ski slope. The ski area itself was trying to attract more summer visitors to make up for the shortfall in winter numbers. Sitting in the café at the bottom of Cairn Gorm's new funicular railway, Cathy, the mountain's ecologist, told me how the company planned to take lifts out at the bottom of the mountain and reinstall just a few of them at the top. 'We are losing infrastructure because we no longer use it – we no longer *can* use it,' she said. 'We will gradually retreat up the hill.'

Few people know more about the history of Scottish snow than Adam Watson, a scientist and naturalist who has spent almost six decades ski-mountaineering and walking around the Cairngorms, studying the habits of ptarmigan and snow bunting, as well as the snow itself. I arranged to meet him at a restaurant near his home in Deeside, a two-hour drive from Aviemore along the high, winding old military road that runs north of the Cairngorms, through Bridge of Brown, Tomintoul and the Lecht. It was Adam's seventy-eighth birthday, and though we had never met I recognized him at once by his long white beard. He spoke in a soft Scots voice, with a scientist's insistence upon impartiality and accuracy. He looked trim and athletic, despite his years, and was dressed in the clothes of a hill walker.

Adam had been raised in the small town of Turriff in Aberdeenshire. He first became interested in snow during a storm when he was seven. He remembers sitting in a room at his home watching it fall in columns from the sky on the roof of a nearby building, seeing it melt and struggle to establish itself, then gradually pile up into a thick layer. When he went outside he was amazed at how the snow had changed everything, how silent it was, and how rich the variety of the crystals.

The following July Adam went with his family to the Cairngorms and was astonished to see lumps of snow on the ground on a distant hill in summer. But it was a year later, on a rainy day in a hotel in Ballater, that the snow and the mountains really caught hold of him. In among a pile of books and magazines that his parents were working through, he found a copy of *The Cairngorm Hills of Scotland* by the naturalist Seton Gordon. Gordon took a great interest in snow, particularly in the patches that lasted all year round. He was intrigued by the plants that grew around the edges of the snow, where it thawed just long enough to allow a short growing season, but he was mainly drawn to the strange phenomenon of the patches themselves, their locations and their longevity. Adam devoured the book. From then on he wanted to spend as much time in the Cairngorms as he could. He wrote to Gordon that autumn, without expecting a reply, and Gordon responded within a week. It was the start of a correspondence about the mountains and the snow that was to last for almost forty years, until Gordon's death in 1977.

Adam began taking his own notes about the snow patches at the age of twelve. At fourteen, he started keeping a diary of snow events and snow levels in the hills. He took up ski-mountaineering at seventeen, and spent most of his winters studying the ptarmigan. When Adam was a postgraduate at Aberdeen University the ptarmigan provided the subject of his research, which conveniently gave him a reason to spend many long days in the mountains in winter. He noted during his field trips how they could fly straight into a snowdrift, kicking snow behind them so that they filled the entrance of the hole and were sheltered from the wind, and how they stayed near enough to the snow's surface that they didn't become buried, but could see when the morning light appeared and when to leave their burrows. 'It takes an Inuit an hour to make an igloo,' Adam said. 'It takes mountaineers an hour or so to make a snow hole. It takes two to three seconds for the ptarmigan.'

By the seventies he had developed a scientific method of snow-patch counting and began a regular survey right across the Highlands. He would start each year on the first of July, repeating the count at the beginning of successive months until the earliest lasting snowfall of the new winter, which usually came in October. In the early days he would climb up to the patches, measuring them at their widest point and marking their locations on large-scale maps. After a while he became used to estimating their size accurately from a distance, often in the snatches of clarity between drifting cloud banks. He knew all the best vantage points for bagging them.

There were two very special patches in the Cairngorms, Adam said: the Pinnacles patch and the Sphinx patch, named after the rock formations above them. These patches both lie in a remote hole in the massif called the Garbh Choire Mor, a few hundred metres south of the summit of Britain's third-highest mountain, Braeriach. They sit in the lee of the cliffs, facing north-east, sheltered from the warm air, the rain and the afternoon sun by the surrounding mountain walls. The snow in the Pinnacles and Sphinx patches lasts longer than any other snow in Britain. The Sphinx patch has melted completely only five times since the mid-1800s. Three of those occasions were after 1995: in 1996, 2003 and 2006.

The mass of data which Adam has accumulated about the snow patches over the years reveals clear patterns. All of the patches in Scotland are disappearing earlier in the season than they once did. 'Since the late 1980s there's been a clear downward trend,' Adam said. The change wasn't statistically significant over the whole period, but if you broke it into two halves it was clear that for the first half of the period, covering the seventies and early eighties, it got snowier, which coincided with the time when the ski areas were doing extremely well. After that there was a strong decline in the sizes and quantities of snow patches. During the twenty-five years up to 1995, the average number of snow patches that survived through the year was 10.7. In the ten years from 1995 to 2005 it was 3.8.

I asked Adam if he considered that Scotland would one day lose its snow altogether. 'It's certainly a concern' — he thought for a moment — 'but snow is always unpredictable.'

As I drove back across the russet and white tops I knew I had to walk into the mountains to see the Sphinx and Pinnacles patches for myself, if I could reach them. From the map it looked to be a day's trek from Aviemore into the dramatic, steep-walled U of the Lairig Ghru, the valley which has been used since ancient times to cross from Strathspey to Deeside, into one of the remotest valleys in the mountains, An Garbh Choire.

I set out with a backpack one sunny morning along the Cairngorm mountain road, turning south by J. Arthur Rank's Coylumbridge Hotel and into the Rothiemurchus forest, where some of the last remnants of the great wood of Caledon which once covered Scotland still grow. The going was flat and firm, and the smell of pine filled the air. After crossing the Allt Mor, whose fast-flowing waters were swollen by the spring thaw, the track climbed towards the mountains among centuries-old trees, between which deep heather and bright green bilberry and juniper soaked up the pools of sunshine that penetrated the forest canopy. Chaffinches sang and flittered between the branches.

After an hour, the trees gave way to heather and black peat, and a mile further on to a far more barren landscape of broken rock. At the entrance of the Lairig Ghru the path crosses a river through a field of giant boulders which have fallen down from the massive scar of Lurcher's Crag 300 metres above. From a distance, the boulder field looks like a bank of shingle, but close up you discover that each large stone weighs several tonnes. The river disappears sporadically beneath porous dams of this built-up debris, reappearing at breaks in the boulders a few hundred metres along.

The path picked a narrow, often invisible route through the rocks. Eventually I lost it altogether and carried on over the lumps of granite at a crawl, sometimes on all fours, concentrating on the stability and size of each stone before trusting it with my weight. Six hundred and fifty metres high I saw my first snow patch by the side of the path. Walking on the patches could be treacherous, I discovered, without knowing how thick each was, or what it might hide. Many were melting from the bottom up, forming streams that gradually hollowed them out.

It took more than an hour to cover the last mile over the rough ground at the top of the pass, and then I could see down between wispy clouds to the savage, empty valley on the far side, where the Lairig Ghru continues on towards Braemar. I tramped through deep banks of snow that lay around the Pools of Dee just below the pass, under a great waterfall carrying the snowmelt off Ben Macdui to the east, then veered west along the great flank of Braeriach into the valley of An Garbh Choire. The going was no easier here. The hillside was again covered with boulders, but these had become overgrown so it was difficult to see where to place each foot. Now and then the terrain was crisscrossed with wide patches of sodden winter grass where the water ran downhill, seemingly without the need of a stream-bed or gully, as if the ground itself were melting.

There were no other humans in this far-flung valley, and no deer, sheep or birds of prey either. Among the few living animals I saw were ptarmigan, and tens of small brown frogs with dark stripes down their backs. They sat half in and half out of the water, which felt to be barely above freezing.

I spent that night halfway up the valley, on the northern bank of the Allt a' Garbh Choire stream which had turned into a rushing, violent torrent. There was a tiny bothy built out of pebbles on the other side, and I wandered for half a mile up and down the riverbank searching for a safe place to jump across to it. When I failed to find one I took my boots off and tried to wade over, but the current was too strong and the water too cold and deep, so I gave up and pitched the tent on a flattish patch of heather. I had pitched the tent in the only flat, rock-free space I could find, which was on the downhill bank of the raging stream, no more than a metre from the swelling water.

As evening fell, I watched the massive wall of Garbh Choire Mor, which translates from Scots Gaelic for 'great, rough corrie', through the canvas doorway. Clouds appeared from behind the ridge, as if the mountain was generating them itself, and swept into the valley, sometimes touching the heather with a finger of mist before hurrying past me and down. Around seven, there was a loud crash of thunder and rain began to fall hard on the tent. The roof dripped. Then another long rumble that began to the southwest, crossed over my head and continued for several seconds as the storm was carried northeast. Next there was lightning, then more thunder. My heart was racing. Could the corrie flash-flood? What if the river simply rose and breached its banks in the night?

I ate some chocolate to calm my anxiety – 'You must eat, food is important,'

Philippe had once told me – and pulled the sleeping-bag drawstring tight. The river churned furiously outside, its sound building now and then into a jet-engine roar before subsiding again. After some time the thunder stopped, the rain eased off, and I relaxed. The heather made a soft bed and I drifted off to sleep.

If the rate of warming in the twenty-first century is unprecedented in recent history, it is slow compared with what geologists believe happened ten thousand years ago, when the great ice sheet that covered most of Scotland thawed. This century, Britain is expected to warm by 3 degrees Celsius; a little over 10,000 years ago, Scotland is thought to have warmed by 7 or 8 degrees in half that time. Chaos ensued: there were catastrophic floods and landslides, as huge rocks and boulders were swept downwards by the force of meltwater.

Some of the effect of this earlier warming can be seen on the walls of the narrow valley of Glen Roy. On one occasion years before, when I was on a skiing visit to Aviemore and the mountain had been closed because of high winds, I had driven west, through Kingussie and Newtonmore, towards Fort William. At Roybridge, I had taken the small track north that wound around fields and over humpback bridges into the steep-sided glen. Like many tourists before, I had been amazed to see three broad, dead-level steps that run all around the valley like tidemarks around a bathtub.

The 'Parallel Roads of Glen Roy', as they are called, run at altitudes of 260, 325 and 350 metres and are so level that for centuries they were used for transport in an area with few tracks. Local people once believed they were built by Fionn MacCumhail, or Fingal, the mythological hunter-warrior, to assist him in hunting deer on horseback when the glen was more wooded. In 1771 the travelling naturalist Thomas Pennant recorded his impression of the 'roads': 'All the description that can be given of the Parallel Roads, or Terrasses, is that the Glen of itself is extremely narrow, and the hills on each side very high, and generally not rocky. In the face of these hills, on both sides of the glen, there are three roads at small distances from each other, and directly opposite on each side. These roads have been measured in the compleatest part of them, and found to be 26 paces of a man five feet ten inches high.' Pennant assumed that they must be man-made. He described them as being 'nearly as exact as drawn with a line of rule and compass', and their length was about seven miles, but there were no traces of buildings or druidical remains that would indicate any religious function. 'The country people think they were designed for the chace [hunt], and that these terrasses were made after the spots were cleared in lines from wood, in order to tempt the animals into the open paths after they were rouzed, in order that they might come within reach of the bowmen, who might conceal themselves in the woods above and below.'

When the travel writer Sir John Carr passed the area in 1807, he claimed that the roads were built by early Scottish kings. A later visitor, the Victorian geologist Thomas Jamieson, wrote of the roads' origins: 'No wonder then that imaginative Highlanders ascribed them to the ancient heroes of their race, and saw in these lines the hunting-roads of Fingal and his companions, that they had made for chasing the deer. In their native language they call them Casan, or "the Footpaths", for on climbing up to them, they are each found to consist of a green ledge, or narrow terrace, jutting out from the face of the hill; so that they actually serve as convenient tracks for walking on.'

By the mid-nineteenth century, the roads of Glen Roy had begun to attract serious scientific attention from the likes of Charles Darwin, the eminent Scottish geologist Charles Lyell and a young Swiss named Louis Agassiz. The remote glen was suddenly thrust into the heart of a very vigorous debate that would reveal one of the great hidden truths about the Earth.

Lyell believed the roads were an ancient maritime shoreline left from a time when Glen Roy had been part of the Scottish coast. His friend Darwin visited the glen in 1838, several decades before he published *On the Origin of Species*. From what Darwin had observed on his journeys in South America, and particularly the marine fossils he had seen in the mountains of Chile, he agreed with Lyell that the Parallel Roads were shorelines, speculating that the reason for them being so far from the coast was that the land had risen.

But it was Agassiz who surmised the real explanation for the roads, two years later. He had been influenced by the work of the European glaciologists Jean de Charpentier, Horace-Bénédict de Saussure and Ignace Venetz, who conjectured from their studies of Alpine valleys that ice was responsible for the apparently bizarre placement of enormous rocks and moraines in the region. Agassiz extended their theories, arguing that the Alps were once covered by a giant ice sheet whose action had created the features of the mountains, and which had then largely melted away. In 1840 he published the hypothesis for the theory of an 'ice age' in the two volumes of *Etudes sur les glaciers*. He travelled to Scotland and Glen Roy the same year, where his theories led him to a different and more radical conclusion than those of Lyell and Darwin. Agassiz recognized in Glen Roy the shape of a valley that had been created or enlarged by a glacier from his fieldwork in the Alps. 'Prepossessed as I was with the idea of glacial agency in times anterior to ours, these phenomena appeared to me under a new aspect,' he wrote later. 'I found the bottom of Glen Spean so worn by glacial action as to leave no doubt in my mind that it must have been the bed of a great glacier.'

Snow falling in the mountains to the north and south of Glen Spean had fed a vast glacier in the valley, he suggested, which had blocked the lower end of Glen Roy and caused it to fill with water, creating the glacial lake whose shores had eroded the valley walls at the level of the highest Parallel Road. Towards the end of the ice age, the giant Glen Spean ice field had retreated to form smaller glaciers, two of which, from Ben Nevis and Loch Treig, had also dammed back the waters, but at a lower level. The Loch Treig ice retreated next, and the lake fell to its lowest level, before finally disappearing completely as the Nevis glacier thawed.

Agassiz went further still. It was not just Glen Roy that had seen such enormous levels of ice. From the examination of other features around the country, he concluded that the ice sheet had once covered the whole of northern Britain. 'It thus appears that the whole range of the Grampians formed a great centre for the distribution of glaciers,' he wrote, 'and that a colossal ice-field spread itself over the whole country, extending in every direction toward the lower lands and the sea-shore.'

Ice-age theory revolutionized the way people understood the landscape, and Scotland's in particular. Between 22,000 and 18,000 years ago, when the massive Scottish ice sheet was at its peak, it was two kilometres deep in places and reached as far south as the Midlands. In the west, dramatic steep-walled glens march towards the sea, row after row. These deep valleys were formed by ice resulting from snow, which fell more heavily in the west, due to the actions of the Gulf Stream.

Glaciers in the west carved out many of the trenches that make up the western Highlands, often following geological fault lines that already existed and had been created millions of years before. The shoreline on this side of the country is carved jagged, with sea lochs formed where the ice ran into the ocean, like the fjords of western Norway.

In the east, where the snowfall and the temperatures were lower, glaciers were fewer and smaller and more inclined to freeze solid to the bedrock, which is why the Cairngorms are more rounded than the Nevis range in the west. But this side of the country still contains some of Scotland's most celebrated glacial landscape features. The Castle Rock in Edinburgh, with cliffs on three sides and a tapering slope down towards Holyrood on the fourth side, was produced by a river of ice moving down towards the Firth of Forth and striking hard igneous rock, and was diverted around it on either side. The crouching lion shape of Arthur's Seat was formed by glaciation, too.

Ice also carved out the Lairig Ghru, and its tributary valley, the An Garbh Choire, where I had camped. The glacier here had begun its journey up at the top of the valley in the Garbh Choire Mor, where the south-westerly winds gathered great mounds of snow that fed the ice, just as they feed the Sphinx and Pinnacles snow patches today.

When Agassiz was proved correct, Darwin was furious. He regretted that he had 'wasted time' on Glen Roy, and described his entanglement with it as the greatest mistake of his career.

I left the tent early the following morning to climb up to the Garbh Choire Mor. The heavens above the great headwall of rock and snow were still busy with rushing grey clouds as I climbed towards my mountaintop encounter with the Sphinx. In ancient Egypt, half-human, half-lion Sphinxes guarded the secrets of the royal tombs; to the Greeks, the Sphinx was a creature who asked riddles and killed those who couldn't answer. In Scotland, the granite cliff-face guarded the snow from the warm south-westerlies.

The ascent was as slow as that of the lower valley as I hopped from rock to rock through the heather and the boulders, and crossed the broad stream-beds towards the ridge. After a mile I reached the corrie's easterly rim, and picked my way through a new field of rocks. Then I was standing on the snow-covered stage of the great natural amphitheatre that had been carved out of Braeriach many millennia before.

From here, the floor of the corrie sloped up steeply to the north before levelling out. I climbed on up across the snow, digging in with my boots to establish a grip on the rounded ice grains beneath my feet. There were signs of cornice-fall avalanches all around, large gaps in the overhanging snow like missing teeth where lumps had broken off and tumbled down the steep walls of the bowl. I moved quickly towards the safety of a patch of rock, guessing that the slope I was climbing was near optimal for an avalanche. Seeds and grass stalks were scattered over the snow, and here and there were insects making their way across it.

When I had reached the rock I stopped to look up at the panorama of cliffs, identifying in turn the Crown Buttress, the Great Gully, the Sphinx Ridge and Pinnacles Buttress, before finally crossing on to the snow wreath beneath the Sphinx.

As well as possessing the unmistakable shape of an Egyptian head-cloth, the Sphinx seemed to have a face, with two lines like chimneys in the rock that began where its eyes would be and ran downwards, defining a nose between them and giving the impression that it was crying. Down beneath her face, in among that stretch of innumerable ice grains, hidden in the protection of the cold dark shadow in this high corrie, was the most persistent snow in Scotland.

Snow was blown here for miles across the Cairngorms, often by ferocious winds which picked up the crystals from the plateau, bashing them together and rolling them along the ground so they broke into fragments. As the snow-bearing winds crossed the top of the Garbh Choire Mor, they formed a vortex in which the snow crystals were tumbled, some of them being pressed on to the cliff-edge to create a cornice while most were sucked down in a swirling, rushing cloud of spindrift into the basin below. Climbers who have stood and watched the snow in these conditions describe it flowing down the cliffs like a waterfall.

I stood and looked at it. Now that I had bagged the patch, I was at a loss as to what to do. I could write my name in the snow, but that seemed like vandalism. So I did what tourists do everywhere: I took some photographs, then gazed at it, trying to absorb its meaning. What struck me then was not the appearance of the snow, but the great noise which I had shut out until then – of rumbling and rushing as water charged invisible beneath my feet, carrying rocks and pebbles along with it. The sound of warming was not a delicate drip-drip, but a cacophony.

Wherever I had travelled, I had asked the people I met the same questions about the changes they saw in the snowfall: had they observed less, and what effect would it have on them? For Billy on Baffin Island, the considerations were practical: without snow covering the rocks where they rode their snow-machines, it was far more difficult to get around; the snow was harder to make into igloos; and the behaviour of animals they lived with changed so that it was harder to hunt. In other Inuit communities, the reduced number of snow patches has left them with fewer places to refrigerate the meat and fish caught in spring and summer. For Philippe and other mountain guides it has affected the length of the winter tourist season, and the local economy. On Mount Rainier, Washingtonians could see their frozen water supply shrinking, and the delicate balance of the snow-dependent animals and plants was shifting. Snow affected so many aspects of life: economies, agriculture, power-generation and ecology. In the American Arctic, a build-up of ice layers in the snowpack produced by the increased melting and freezing have led to a catastrophic reduction in Peary caribou, which have been prevented from digging down to eat the lichens beneath. Snow-cover also affects the Earth's albedo: snow reflects the sun's rays and so helps to reduce the planet's warmth; with less snow, the world is warming up more quickly.

In Scotland, apart from winter sports, there was a cultural connection with snow that seemed to me to be at risk. Snow has been woven into Highland folklore and history. It was said that the Munro clan on Ben Wyvis and Ben Nevis was able to have the land for free as long as the chiefs could produce a snowball for the Scottish king in midsummer. There was no snow on Ben Wyvis in autumn now. Legend had it that the absence of snow on White Mounth in the Lochnagar range foretold of doom. 'When ye White Mounth frae snaw is clear, Ye day of doom is drawing near,' ran the rhyme. White Mounth had been increasingly clear in recent years. In the 1800s and early 1900s a bank of snow called the Cuidhe Crom, or curved wreath, on Cairn Gorm signified the start of the agricultural season in the area: if it broke in two before Midsummer Day, it was evidence of an early growing season, and time to take the sheep and cattle up the hill. Now the snow was less reliable, and farmers' incomes were not dependent on such knowledge.

One of snow's most famous historical interventions in Scotland occurred on the night of the massacre at Glencoe in 1692, when Captain Robert Campbell of Glenlyon returned the hospitality of the Glencoe MacDonalds by murdering thirty-eight of them, including women and children, during a blizzard. A short version of the story starts with MacIan, clan chief of the MacDonalds and a man with the reputation of a mafia don, arriving at Fort William to swear an oath to King William III on 30 December 1691, one day before the deadline for swearing allegiance expired. MacIan was told that he had come to the wrong place, and was sent instead to Inverary, seventy-four miles away across the mountains, where Sir Colin Campbell would take his oath. He could not possibly make it in time, but perhaps conscious of the political capital his many enemies would make if he failed to show willing, MacIan set out. The country was covered with deep snowdrifts, which meant he didn't reach Inverary until the second of January. After some further delay, MacIan's oath, by now several days late, was initially accepted, then discarded, and the Scottish Secretary and Master of Stair, Sir John Dalrymple, was given the excuse he needed to order murderous retribution on the MacDonalds, of whose thieving and banditry he was determined to make an example.

Glenlyon and 120 soldiers were despatched to carry out the killing. They were put up by the unsuspecting MacDonalds for twelve days and nights, during which they ate and drank the clan's winter supplies. A snowstorm blew up in the early hours of 13 February 1692, when the attack had been scheduled, the flakes coming thick and fast enough to muffle the shouts, presumably of warning, from guilt-ridden government officers to MacIan's eldest son, John. In his *History of England*, the nineteenth-century historian Thomas Babington Macaulay noted that the snowstorm was enough to delay Lieutenant-Colonel Hamilton, who was marching from Fort William with troops to assist Glenlyon and to block the exits from the valley, by several hours. At 5 a.m., Campbell's men fell on their hosts, under orders to 'put all to the sword under 70', but some of them, including John and his family, escaped into the snowstorm, which helped hide them and covered their tracks. Old MacIan was shot dead, his wife dragged out into the snow, stripped and left to die, which she did a day later. The women, children and old people who were not killed were burned out of their homes and left to make their way over the mountains to safety in the snow. During the journey many of them died of exposure.

In his account of the incident, Sir Walter Scott related how the blizzard helped save those who ran from the Campbells: 'The great fall of snow, which proved fatal to several of the fugitives, was the means of saving the remnant that escaped.' In Scott's version, snow – the killer and redeemer – is a projection of the landscape, of Highland nature itself. This was a theme he returned to often in his writings, using snow to portray mood, and as a metaphor in fictional works such as *Guy Mannering* as well as in his histories.

Other Scottish poets, writers and artists have used snow to evoke the spirit of this northern country, including James Macpherson and Robert Burns, who writes in *Tam O'Shanter*:

> But pleasures are like poppies spread,
> You seize the flower, its bloom is shed;
> Or like the snow falls in the river,
> A moment white – then melts for ever.

Walter Scott's portraitist, Sir Henry Raeburn, used snow and ice as the backdrop for what has become Scotland's most famous work of art, *The Reverend Robert Walker Skating on Duddingston Loch* (1795), in which the minister of the Canongate Kirk scoots one-legged across the frozen water with a technique that would have won admiring glances from fellow members of the Edinburgh Skating Society. Frozen Duddingston Loch was a regular sporting venue for the people of Edinburgh: it was also used for curling, a Winter Olympic discipline that was developed in Scotland and played on specially constructed ponds across the Highlands. It isn't often played that way any more: the curling ponds in the Highland villages rarely freeze today.

Scotland's idea of itself, it seemed to me as I stood in the Garbh Choire, was of a place that was cold. But perhaps I was merely in the grip of *nostalgie de la neige*, romanticizing Highland history, as so many had done before. Would it matter, I had asked Adam, if there was less snow in Scotland? His answer was abrupt. The average person didn't really give a hang about it, he said, with some regret. Most people saw snow as a nuisance which got in the way of the commute to work. Hardly anyone cared when the white wreath on Cairn Gorm broke any more because the people had become detached from the land, even the farmers.

And what about him? We sat in silence for a short moment as he took a swig of his coffee. 'It will be interesting, whatever happens,' Adam said, with the hint of a smile. 'But I'm not going to be around to see it.'

A bank of cloud half-filled the valley below me. I had watched it stream in and upwards from the eastern end of the Lairig Ghru and now only the upper parts of the mountains stood above it. From this distance it could have been mistaken for a white sea of glacier ice which left only the high mountains sticking up, like islands.

It was hot work crossing the heather-covered boulders again, and after I had hiked across the white slope back towards the valley I realized I had run out of water. I stopped by a stream that ran down from a snow patch. The water ran clear as glass over a gravel bed, with a tinkling from the tiny overflowing pools all around. I scooped the liquid up thirstily and it ran down my chin. I could not remember drinking anything that was so delicious, yet which tasted so much of nothing. There was only the sensation of swallowing the undiluted cold of this new water.

When scientists examine other planets to see if they could ever have sustained life, they look for water. If there is water, there is the chance of living.

I set off down into the valley again. It was a long way to Aviemore and my back was already aching from the weight of my pack, but I felt exuberant to be out walking, without a path, alone in this great valley, among these granite hills.

Halfway down An Garbh Choire a burst of flapping broke out at my feet and a pair of ptarmigan shot out of the heather with whirring wings and a long croaking call. They wore their camouflaged spring plumage, white for the snow and stone grey for granite. I watched them fly across the heather into the distance for a moment, before losing them against the snow patches on the shoulder of Braeriach.

CHAPTER II

It Will Be Spring Soon

London

I am sitting by the upstairs window during a late winter storm watching rain fall in the road below. It is a gusty day, and I am wondering if one of the poor-fitting aluminium windows that we haven't got around to replacing might blow in. Down the stairs I can hear Lucy singing 'Pitter Patter Raindrops' while she dresses Eddie, whom I recall we named while driving around New York State more than two years before. 'I'm wet through,' sings Lucy, 'So are you.' Eddie's giggles bubble along the corridor and make me smile.

We only had one snowfall in London this winter, which I suppose is about average. Now the rising temperatures and increasing rainfall herald the arrival of the new season. Snow, in the northern hemisphere at least, is melting again. Scottish skiing is over for the year. In Paradise, the shoots of avalanche lilies will soon be poking through the snowfields that linger in the high alpine pastures, the marmots are blearily waking up and finding the empty nest next to them where a pika had spent the winter. On Thompson Pass, the wet clumps of snow will be falling from the trees where the moose gather, and when Matt goes out on the pass he will take his gun, in case he meets a half-asleep bear. Billy will be greeting the first clients of the year at Pangnirtung airport, and taking them out on the land. The ski patrollers of Baker will be thinking about closing down the resort, saying their goodbyes and contemplating summer jobs in forestry or on the coast. The frozen world is in retreat.

The nature writers and poets have loved this time of year, when the earth bursts back to life. If summer is a time already partially shadowed by thoughts of the coming autumn, then the last days of winter, when the sun returns to the sky, are the moment to anticipate the new spring. Robert Frost wrote: 'I know that winter death has never tried / The earth but it has failed: the snow may heap / In long storms an undrifted four feet deep / As measured against maple, birch and oak, / It cannot check the peeper's silver croak.'

Others have found in spring a metaphor for love: 'Over the winter glaciers, / I see the summer glow,' wrote Ralph Waldo Emerson, 'And through the wild-piled snowdrift / The warm rose buds below.' Henry David Thoreau, meanwhile, remarked on the 'crisis' of spring: 'The change from storm and winter to serene and mild weather, from dark and sluggish hours to bright and elastic ones, is a memorable crisis which all things proclaim.'

My feelings about the end of winter are mixed. When I was a child, leaving the snow behind was always for me the worst part of the holiday, accompanied as it was by a sense of returning to reality and responsibility. The drive down from the mountains was usually accompanied by heavy silence. But when a thaw sets in there is no point in wanting the snow to remain. Like a failed love affair that drags painfully on, you wish it would go more quickly and leave you to work with its memory. The melting bergs with rounded edges linger on the ground as the rain comes down, the white on the grass ebbing to leave acres of black and green mud. In the dirty city, everything that the snow covered is revealed again, along with an extra layer of rubbish and cigarette butts for good measure. The snow covers what lies beneath. It doesn't remove it. In the mountains, snow throws up its victims in the spring.

My snow tour ended shortly after my fortieth birthday. My own 'memorable crisis' fitted between the arrival of children and the turning of one season of life into another. I no longer feel the desperate need to travel; in fact, I rather want to take a rest from it and be with my family. I have had enough of freedom for a while, though I know I can find it again if I need it.

During this tour I found my own terrible corner of winter as well as paradise, churning vortices as well as chocolate-box visions of crisp white perfection. Snow has many facets; Monet's and Turner's interpretations both show versions of the truth, but Bruegel, I believe, painted the fairest portrait of winter. There are cold winds and poor seasons for hunting, but the world is also more beautiful and there are people enjoying themselves, playing games out on the ice.

We are not going to move to the country, you may have guessed. I still long for open space, but have also been persuaded of the benefits of life in the city, its art galleries and its culture and its libraries, and the connections to be made between them. I have followed Philippe's advice and signed up for a class at the climbing centre up the road, so that I can become proficient with ropes and carabiners and practise being up high. Nick has a plan for us to climb Mont Blanc, and maybe one day we will.

Meanwhile, I will try to pass on my love of snow to our boys, although I recognize that they may not respond to it the way I did. It is strange to think that my own love for it began with a single photograph.

In the top room, I dig out the picture of my father in his racing bib, with the Austrian Alps behind him, and his happy laugh. Being a father is about being there, I decide; if you don't manage that, you haven't even begun, and now I need to be around for my children. We are not the same people, my father and I, just as my sons are not the same as me. His blood is mixed in me with my mother's, and she is very strong in ways that he was not.

I can hear Lucy calling now from downstairs, and I shout a reply. Out of the window I see she is trying to pack the boys into the car. The tops of their three blond heads are milling around in all the wrong ways, each going in his own direction. As I look at them from above, a tremendous warmth wells up in me and my eyes grow wet. It seems to be a function of getting older, this urge to cry.

I put the picture of my father away in the desk drawer, a little further back than it needs to be. It's time to go. We are off to buy some skiing clothes. We have booked a late deal to the Alps. The snow will be patchy and slushy, but there may be just enough left. In no time, I am sure, the boys will be flying down the slopes, oblivious to the rest of the world.

In the nineteenth century, Fritjof Nansen wrote that skiing washes civilization clean from our minds by dint of its exhilarating physicality. By extension, I believe that snow helps strip away the things that don't matter. It leaves us thinking of little else but the greatness of nature, the place of our souls within it, and the dazzling whiteness that lies ahead.

A Snow Handbook

How snow is created

1. A cloud forms when moist air is cooled, usually by being forced upwards, and water vapour condenses out into droplets.

2. If the cloud is cold enough, some of the moisture in it turns into ice, which forms on tiny dust particles that blow around in the atmosphere. These microscopic ice nuclei are the seeds of snow crystals.

3. Water molecules migrate from droplets and attach to the freezing nuclei, growing ice crystals, or snow. Most of these crystals have a hexagonal symmetry.

4. When a snow crystal grows large enough, it starts to fall towards the earth. Some take hours to reach the ground. Many never get there, but melt or evaporate on the way down.

Snow-crystal types

The shapes of snow crystals are determined by the temperatures and pressures they are subjected to. Snowflakes are all different because each crystal's temperature and pressure history is different. Crystals have, however, been classified into certain types. Some methods of snow classification identify more than a hundred basic crystal shapes. Here are a few:

1. Needle

2. Column

3. Hexagonal plate

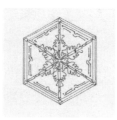

4. Crystal with broad branches

5. Simple stellar
crystal

6. Fernlike crystal

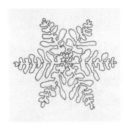

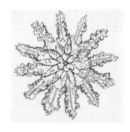

7. Plate with dendritic
extensions

8. Dendritic crystal
with 12 branches

'Announced by all the trumpets of the sky,
Arrives the snow, and, driving o'er the fields,
Seems nowhere to alight: the whited air
Hides hills and woods, the river, and the heaven,
And veils the farmhouse at the garden's end.
The sled and traveller stopped, the courier's feet
Delayed, all friends shut out, the housemates sit
Around the radiant fireplace, enclosed
In a tumultuous privacy of storm.'
      Ralph Waldo Emerson, 'The Snow-Storm'

THE LARGEST EVER SNOWFLAKE WAS FOUND
AT FORT KEOGH, MONTANA, USA, ON
27 JANUARY 1887 – IT WAS 15 INCHES
(38CM) WIDE BY 8 INCHES (20CM) THICK.

10 snowy places

Thompson Pass, Valdez, Alaska, USA

Achishko, Western Great Caucasus, Russia

Klyuchevskaya mountains, Kamchatka, Russia

Mount Baker and Mount Rainier, Cascades, Washington State, USA

Revelstoke, British Columbia, Canada

Silver Lake, Colorado, USA

Bessans, France

Mount Hutt, New Zealand

Jostedal glacier, Norway

Niseko, Japan

'The North wind doth blow and we shall have snow,
And what will poor robin do then, poor thing?
He'll sit in a barn and keep himself warm
and hide his head under his wing, poor thing.'

Nursery rhyme

Michelangelo's snowman

Apprentices at artists' studios in Renaissance Florence traditionally carved snow sculptures every time it snowed. Following a snowstorm in Florence on 20 January 1494, the eighteen-year-old Michelangelo was instructed by his patron Piero de'Medici to create a large figure out of snow in the courtyard of the Medici Palace. Historians have interpreted the commission as an insult to the artist, which helped turn Michelangelo against the Medicis. We can only imagine what the young man did with his ephemeral material. Did he build the most lifelike snowman ever?

10 snow stories

A Christmas Carol by Charles Dickens (1843)

The Snow Queen by Hans Christian Andersen (1845)

The Call of the Wild by Jack London (1903)

The Lion, the Witch and the Wardrobe by C. S. Lewis (1950)

The Worst Journey in the World by Apsley Cherry-Garrard (1951)

The Snowman by Raymond Briggs (1978)

Miss Smilla's Feeling for Snow by Peter Hoeg (1994)

Northern Lights by Philip Pullman (1995)

Snow Falling on Cedars by David Guterson (1995)

South: The Endurance Expedition by Ernest Shackleton (1999)

How avalanches occur

Snow can avalanche in several different ways. These are the main types:

Slab avalanche

Caused when a cohesive slab of snow breaks away from the snowpack due to a weak, badly bonded layer beneath. Slab avalanches can range in width from a few yards to over a mile, but typically are the size of half a football pitch and 1−2 feet in depth.

Loose snow avalanche

Sometimes called sluffs, these usually start from a single point and fan outwards over the upper layers of snow as they descend. They tend to be smaller than slab avalanches.

Ice avalanche

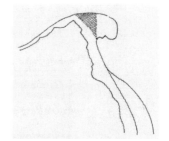

Occurs when a block of ice detaches from a steep or overhanging cliff or glacier, gathering snow or rock as it falls.

Cornice fall avalanche

Happens when wind-blown snow breaks from a ridgeline or cliff edge, often triggering further avalanches in its descent.

Glide avalanche

A wet snow avalanche that occurs when warmer temperatures or rain have weakened the snowpack, causing the whole to slide down the slope, exposing the hill beneath.

Slush avalanche

Occurs on very gentle slopes, when a snowpack becomes saturated quickly and breaks up, releasing slabs of wet snow.

'*This first fallen snow*
is barely enough to bend
the jonquil leaves.'
Matsuo Bashō (1644–94),
Japanese poet

THE GREATEST RECORDED SINGLE
SNOWFALL WAS 189 INCHES (480CM) IN
MOUNT SHASTA SKI BOWL, CALIFORNIA,
USA, ON 13–19 FEBRUARY 1959.

How to survive an avalanche

The best way to avoid being killed by an avalanche is not to get caught in one. Recognizing the climate, terrain and snow conditions that create avalanches is something of an art form, and there are many courses and books that instruct winter sports enthusiasts on the subject (see Bibliography). If you do become caught up in an avalanche, there are a few ways in which you can marginally increase your chances of survival.

1. Always wear an avalanche transceiver and do not ski off-piste alone.

2. When the avalanche releases, shout to attract attention and try to escape its path.

3. Try to jettison your skis or board but keep your pack on – this provides padding and it may help you to float to the surface.

4. Try to 'swim' as hard as possible so as to remain on the surface of the snow while it is moving.

5. As the avalanche slows, use your hand to try to keep an air space around your mouth.

6. If you are completely buried when the avalanche has stopped, relax. You are unlikely to be able to move and it is better to conserve energy and wait for rescue.

'*I used to be Snow White,*
but I drifted.'

Mae West

10 snow movies

Nanook of the North (1922, dir. Robert J. Flaherty)

It's a Wonderful Life (1946, dir. Frank Capra)

The Heroes of Telemark (1965, dir. Anthony Mann)

Doctor Zhivago (1965, dir. David Lean)

Downhill Racer (1969, dir. Michael Ritchie)

The Shining (1980, dir. Stanley Kubrick)

Fargo (1996, dir. Joel Coen)

Atanarjuat the Fast Runner (2001, dir. Zacharias Kunuk)

Touching the Void (2003, dir. Kevin Macdonald)

Eight Below (2006, dir. Frank Marshall)

Snow for the king

Snow has occasionally been used to curry favour with royalty, and never is it more valuable than in midsummer. In Scotland, the Munro clan chiefs were allowed to occupy land on Ben Wyvis provided that they presented the king with a midsummer snowball, brought from one of the long-lasting snow patches on the mountain. In medieval Japan, the feudal lord of Akimoto won the approval of the shogun by bringing midsummer snow from Mount Fuji. He started his journey with porters and horses carrying a great load of snow, and by the time he reached Edo (now Tokyo) there was just enough left to ice drinks and make snow cones.

'Crossing half the sky,
on my way to the capital,
big clouds promise snow.'
Matsuo Bashō (1644–94),
Japanese poet

10 countries with significant avalanche fatalities

France 30 deaths average per year

Austria 26

Switzerland 23

USA 19

Italy 19

Canada 10

Norway 5

Spain 4

Slovakia 3

Poland 2

THE WORLD'S BIGGEST SNOW CASTLE IS
BUILT IN KEMI, FINLAND, EVERY WINTER.

10 strange snowfalls

Pink snow fell at Durango, Colorado on 9 January 1932.

Red snow fell in the Alps on 14 October 1775 and 3−4 February 1852.

Brown snow fell at Mount Hotham, Victoria, Australia, in July 1935.

Black snow fell in the Lewis, Herkimer, Franklin and Essex counties of New York in April 1889.

Blue snow fell in western New York State in January 1955.

Yellow snow fell in South Bethlehem, Pennsylvania, on 16 March 1879.

Orange snow fell in Omsk, Russia, on 31 January 2007.

Snow fell in the Sahara desert in southern Algeria on 18 February 1979.

Snow fell at Oahu, Hawaii, on 3 March 1953.

In the UK, snow fell in June 1975 in Buxton, Derbyshire, Birmingham, Edinburgh, Newark, Grantham, Peterborough, Cambridge and Colchester. This was the latest snowfall recorded in the UK since the late nineteenth century.

THE WORLD RECORD FOR THE LARGEST SNOW SCULPTURE WAS ESTABLISHED AT THE HARBIN SNOW AND ICE FESTIVAL IN CHINA IN 2007. THE COMPOSITION (DEPICTING THE NIAGARA FALLS AND THE BERING STRAIT) WAS 250 METRES LONG, NINE METRES HIGH, AND USED OVER 13,000 CUBIC METRES OF SNOW.

How to build an igloo

1. Choose a site near a good supply of wind-compacted snow and mark out a circle in the snow.

2. Tramp the snow down around the circle to make a flat foundation free of lumps.

3. Cut blocks, angling their edges so that they lean inwards, and build up in a spiral, shaping each block with a knife or saw to ensure a good fit.

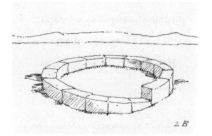

4. When the igloo reaches shoulder-height, cut out a small entrance so that you can continue to build from the inside.

5. Exaggerate the lean on the blocks towards the top.

The last block to place is the central keystone, which must be lifted up from inside, tilted sideways to get through the hole in the roof.

6. Pack any spaces between the blocks with loose snow and smooth the walls over.

7. Vacate the igloo, leaving a lighted candle or stove to make it airtight, then pierce a small hole in the roof for ventilation.

'Polar exploration is at once the cleanest and most isolated way of having a bad time which has been devised.'
Apsley Cherry-Garrard

10 well-adapted animals

arctic fox	lemming
polar bear	ptarmigan
arctic hare	caribou
lynx	husky dogs
musk ox	marmot

10 glacier facts

❄ Glaciers are created by snow that doesn't melt but changes into dense ice.

❄ One-tenth of the Earth's land is currently covered by glaciers.

❄ At the height of the last ice age, around 20,000 years ago, glaciers covered almost a third of the land.

❄ Three-quarters of the world's fresh water is frozen in glaciers.

❄ In places, the Antarctic ice sheet is more than four kilometres thick.

❄ If all the ice on land thawed, the sea level would rise by around 70 metres.

❄ The Antarctic ice sheet is at least 40 million years old.

❄ Ice crystals within glaciers can grow to the size of a cricket ball.

❄ The fastest glacier whose speed has been clocked dashed 12 kilometres in three months.

❄ The weight of ice in parts of Antarctica is thought to have depressed the land up to 2.5 kilometres below sea level.

> '*Their wintry garment of unsullied snow*
> *The mountains have put on...*'
> Robert Southey,
> 'The Poet's Pilgrimage'

ONE OF THE EARLIEST DEPICTIONS OF SKIING IS
A ROCK CARVING IN RØDØY, NORWAY, WHICH IS
THOUGHT TO BE AT LEAST 4,000 YEARS OLD.

The Great Eskimo Vocabulary Hoax

It is often claimed that the Eskimo have fifty, a hundred or even five hundred words for snow. This idea originated with the anthropologist Franz Boas who cited four different 'Eskimo' words for the English 'snow', and the number inflated steadily over the years. In 1986, the anthropologist Laura Martin attempted to puncture the myth, arguing that Eskimo is not a single language and that the very concept of words in Eskimo languages is meaningless, as their 'words' are collections of elements which are analogous to sentences in English. Martin held that West Greenlandic had two distinct linguistic roots for snow. Later, Geoffrey K. Pullum wrote a humorous article, 'The Great Eskimo Vocabulary Hoax', which appeared in his book of the same title (1988), poking fun at those who propagated the myth. However, it now appears that both Martin and Pullum overstated their case. There was never any intention to deceive – no 'hoax' – and most 'Eskimo' languages do have more than the two roots Martin claimed, however you count them. Here are just a few of the snow-related words in Central Alaskan Yup'ik:

Qanuk	Snowflake
Qanir	To snow
Kaneq	Frost
Kanevvluk	Fine snow/rain particles
Natquik	Drifting snow

Nevluk	Clinging debris
Aniu	Snow on ground
Apun	Snow on ground
Qanikcaq	Snow on ground
Muruaneq	Soft deep snow
Qetrar	For snow to crust
Qerretrar	For snow to crust
Nutaryuk	Fresh snow
Qanisqineq	Snow floating on water
Qengaruk	Snowbank
Utvak	Snow carved in block
Navcaq	Snow cornice; snow (formation) about to collapse
Navcite	To get caught in an avalanche
Pirta	Blizzard
Pirtuk	Blizzard
Cellallir	To snow heavily
Pirrelvag	To blizzard severely

'*If you march your Winter Journeys you will have your reward,*
so long as all you want is a penguin's egg.'

Apsley Cherry-Garrard

10 snow artworks

February in *Les Très Riches Heures du Duc de Berri*
by the Limbourg Brothers, 1412–16

The Hunters in the Snow by Pieter Bruegel, 1565

The Reverend Robert Walker Skating on Duddingston Loch
by Sir Henry Raeburn, 1795

Winter Landscape by Caspar David Friedrich, 1811

Snow Storm: Steamboat off a Harbour's Mouth
by J. M. W. Turner, 1842

Women in the Snow at Fujisawa by Utagawa Hiroshige, 1851–2

The Magpie, Snow Effect, Outskirts of Honfleur by Claude Monet, 1868–9

Rooftops under Snow by Gustave Caillebotte, 1878

Giant Snowball by Andy Goldsworthy, 2000

Embankment by Rachel Whiteread, 2005

'I love snow, and all the forms
Of the radiant frost...'
Percy Bysshe Shelley,
'Invocation'

The Abominable Snowman

The term was first coined by the columnist Henry Newman in 1920, following the return from Mount Everest of a reconnaissance party led by Lieut.-Col. C. K. Howard-Bury. The expedition had observed dark forms moving in the snowfields high above their camp on the Tibetan side and found footprints 'three times those of normal humans'. The sherpas probably told Howard-Bury that the footprints had been made by *meh-teh kangmi*, meaning 'man-sized wild snow creature', but in his report he wrote 'Metoh-Kangmi', and by the time the telegraphist had cabled this to India, *metoh* had become *metch*. Newman, who had the reputation of being an expert on Tibetan dialects, was asked to assist with the phrase and came up with the preposterous translation 'Abominable Snowman'. Needless to say, the phrase launched a thousand newspaper articles and several expeditions to find the creature, but the *meh-teh* or yeti proved elusive.

THE WINTER OF 1962/63 WAS THE COLDEST IN THE UK
SINCE 1740 AND BECAME KNOWN AS 'THE BIG FREEZE'.

'*It is better to go skiing and think of God, than go to church and think of sport.*'
Fridtjof Nansen

'*In the bleak midwinter,*
Frosty wind made moan,
Earth stood still as iron,
Water like a stone;
Snow had fallen, snow on snow,
Snow on snow,
In the bleak midwinter,
Long ago.'
Christina G. Rossetti,
'A Christmas Carol'

Glossary

Ablation	Reduction in the quantity of snow or ice by processes such as melting, evaporation or sublimation.
Alpine skiing	Skiing downhill with bindings fixed at the toe and heel.
Aspect	Direction in which a snow-covered slope faces.
Avalanche air bag	A bag worn by off-piste skiers that can be inflated when an avalanche strikes to keep the victim near the snow's surface.
Beacon	Electronic device used to locate avalanche victims.
Bed surface	Snow surface on which an avalanche slides.
Bergschrund	Often deep crevasse that forms between rock and the snow and ice that is flowing away from it at the head of a glacier.
Blizzard	A snowstorm with strong winds, low temperatures and poor visibility.
Chinook	A warm down-slope wind (see *föhn wind*).
Continental climate	A climate found in the interior of a large land mass which in winter tends to be cold and dry.

Corn snow	Snow with large, round-grained crystals formed by melting and freezing.
Cornice	An overhang of snow formed by the wind, most commonly on the lee side of a ridge.
Couloir	A steep snow- or ice-filled chute in a rock wall.
Crevasse	A deep crevice or fissure, as in a glacier.
Crud	Heavy, sticky snow on which it is difficult to ski.
Crust	Of snow, a hard, cohesive layer that sits on a softer layer.
Cryosphere	Those parts of the Earth where water is frozen all the year round.
Dendrite	A snow crystal with complex, fernlike branches.
Depth hoar	Large, fragile cup-shaped crystals of ice that grow within the snowpack and frequently cause avalanches. A type of faceted snow (q.v.).
Dry avalanche	An avalanche that occurs in subzero temperatures.
Faceted snow	Snow with large crystals and poor bonding that forms when there are large temperature gradients in the snowpack.
Firn	Old snow that has become granular and dense, an intermediate stage between fresh snow and glacier ice.
Firnspiegel	A thin, highly reflective sheet of clear ice that forms on the surface of spring snow.
Föhn wind	A warm, dry wind that flows downhill from a mountain, rapidly melting the snow. The *föhn* (also *foehn*) is an Alpine term; it is also known as a *zonda* (Argentine Andes), *chinook* (Rocky Mountains) and *halny wiatr* (Poland).

Glide	Movement of the whole snowpack together.
Graupel	Soft pellets of heavily rimed snow particles.
Hoarfrost	A feathery deposition of ice crystals that can form on objects at temperatures below freezing.
Kamik	Traditional Inuit sealskin or caribou-hide boot.
Langlauf	A form of free-heel skiing, also known as Nordic skiing, cross-country skiing and *ski de fond*.
Leeward / lee slope	A slope that is sheltered from the wind, typically by a ridge. Such slopes can be dangerous because snow loads up on them.
Maritime climate	A warm, wet climate typical of islands and coastal regions. Sometimes called a marine climate.
Metamorphism	Process by which snow changes in structure and texture in the snowpack.
Permafrost	Permanently frozen ground.
Powder snow	Freshly fallen soft snow.
Probe	Long, usually foldable rod used to feel for avalanche victims.
Propagation	The growth of a crack, as in a snowpack.
Qomatiq	Inuit gear sled.
Qaggi	Inuit public building, usually in the form of a large ice dome, for music or ceremonial gatherings.
Quinzhee	A shelter formed by hollowing settled snow.
Rime	A deposit of ice formed when super-cooled water droplets touch an object.

Run-out	The end of an avalanche path, where the avalanche leaves its debris.
Sastrugi	Sharp, irregular ridges on the snow formed by the wind.
Settlement	The process by which snow on the ground becomes more dense due to gravity.
Snowboard	(a) A flat white plank used to measure snowfall depth; (b) a flattish plank with bindings that can be used to travel downhill.
Snow flurry	A light, short-lived snow shower.
Snowline	The lower edge of a snow-covered area.
Snowmobile or snow-machine	Vehicle designed for crossing snow, often driven by a broad rubber track and steered with skis.
Sublimation	Process by which ice changes to water vapour without passing through a liquid state, and the reverse process.
Super-saturation	State in which humidity is greater than 100 per cent.
Telemark skiing	A type of 'free-wheel' downhill skiing using the Telemark turn devised by Sondre Norheim.
Umiak	A large Inuit boat used for transporting people or goods or for whaling.
Weak layer	A plane or lamina of poorly bonded or fragile snow crystals within the snowpack, which is likely to break.
Whoompf/ whumpfing	The sound of a snowpack collapsing on a weak layer within the pack.
Wind slab	Crust of hard-packed snow formed as it is deposited by the wind.

Bibliography

Agassiz, Louis, *Geological Sketches*, Ticknor & Fields, Boston, 1866

Agrawala, Shardul, *Climate Change in the European Alps: Adapting Winter Tourism and Natural Hazards Management*, OECD Publishing, 2007

Ball, Philip, *H₂O: A Biography of Water*, Weidenfeld & Nicolson, London, 1999

Bentley, W. A., and Humphreys, W. J., *Snow Crystals*, McGraw-Hill Book Co., New York and London, 1931

Bilby, Julian W., *Among Unknown Eskimo*, Seeley Service, London, 1923

Blanchard, Duncan C., *The Snowflake Man: A Biography of Wilson A. Bentley*, McDonald & Woodward Publishing Co., Blacksburg, VA, 1998

Boas, Franz, *The Eskimo of Baffin Land and Hudson Bay*, AMS Press, New York, 1975

Burroughs, W. J., 'Winter Landscapes and Climatic Change', *Weather*, 1981

Cable, Mary, *The Blizzard of '88*, Macmillan, New York, 1988

Carr, Sir John, *Caledonian Sketches*, Mathews & Leigh, London, 1809

Cherry-Garrard, Apsley, *The Worst Journey in the World: Antarctic 1910–1913*, Constable, London, 1922

Coolidge, W. E. B., *Josias Simler et les origines de l'Alpinisme jusqu'en 1600*, Glénat, Grenoble, 1989

Doesken, Nolan J., and Judson, Arthur, *The Snow Booklet: A Guide to the Science, Climatology and Measurement of Snow in the United States*, Colorado State University, Fort Collins, CO, 1996

Doughton, Sandi, 'Life on Rainier', The News Tribune, Tacoma, WA, 25 October 2007

Fagan, Brian M., *The Little Ice Age: How Climate Made History 1300–1850*, Basic Books, Boulder, CO, 2000

Fleming, Fergus, *Killing Dragons: The Conquest of the Alps*, Granta, London, 2000

Fraser, Colin, *Avalanches and Snow Safety*, John Murray, London, 1978

Fredston, Jill, *Snowstruck: In the Grip of Avalanches*, Harcourt, New York, 2005

Fredston, Jill, and Fesler, Doug, *Snow Sense: A Guide to Evaluating Snow Avalanche Hazard*, Alaska Mountain Safety Center, Anchorage, AK, 1994

Gribble, Francis H., *The Early Mountaineers*, Fisher Unwin, London, 1899

Hooke, Robert, *Micrographia, Or Some Physiological Descriptions of Minute Bodies Made by Magnifying Glasses, With Observations and Inquiries Thereupon*, I. Martyn and J. Allestry, London, 1665

Hulme, Mike, *Climate Change Scenarios for the United Kingdom: The UKCIP Briefing Report*, Tyndall Centre for Climate Change Research, School of Environmental Sciences, Exeter, 2002

Jacobson, Steven A., *Yup'ik Eskimo Dictionary*, Alaska Native Language Center, University of Alaska, Fairbanks, AK, 1984

Jamieson, Thomas F., 'On the Parallel Roads of Glen Roy', *Quarterly Journal of the Geological Society*, 1863

Kepler, Johannes, *The Six-cornered Snowflake, edited and translated by Colin Hardie. With Essays by L. L. Whyte and B. J. Mason*, Clarendon Press, Oxford, 1966

Kinney, Matt, *Alaska Backcountry Skiing: Valdez and Thompson Pass*, Prince William Sound Books, Valdez, AK, 2006

Kirk, Ruth, *Snow*, University of Washington Press, Seattle, 1998

LaChapelle, Edward R., *Secrets of the Snow: Visual Clues to Avalanche and Ski Conditions*, University of Washington Press, Seattle, 2001

LaChapelle, Edward R., *Field Guide to Snow Crystals*, University of Washington Press, Seattle, 2001

Lamb, H. H., *Climate, History and the Modern World*, Methuen, London, 1982

Langmuir, Eric, *Mountaincraft and Leadership*, Mountain Leader Training Board, Manchester, 2003

Lethcoe, Jim and Nancy, *Valdez Gold Rush Trails of 1898–99*, Prince William Sound Books, Valdez, AK, 1996

Libbrecht, Kenneth, and Rasmussen, Patricia, *The Snowflake: Winter's Secret Beauty*, Colin Baxter Photography, Grantown-on-Spey, 2004

London, Jack, *The Call of the Wild*, Macmillan, New York, 1903

Lopez, Barry, *Arctic Dreams: Imagination and Desire in a Northern Landscape*, Macmillan, London, 1986

Lunn, Arnold, *A History of Ski-ing*, Oxford University Press, Oxford, 1927

Lunn, Peter, *The Guinness Book of Skiing*, Guinness Superlatives, London, 1983

McGhee, Robert, *Ancient People of the Arctic*, UBC Press in Association with the Canadian Museum of Civilization, Vancouver, 1996

McGhee, Robert, *The Last Imaginary Place: A Human History of the Arctic World*, Oxford University Press, Oxford, 2006

Moffett, Charles S. et al., *Impressionists in Winter: Effets de Neige*, Philip Wilson, London, 1998

Mullett, Mary B., 'The Snowflake Man', *The American Magazine*, February, 1925

Nansen, Fridtjof, *The First Crossing of Greenland*, Longmans, London, 1890

Owingayak, David, *Arctic Survival Book: Safety on Land, Sea and Ice*, Inuit Cultural Institute, Eskimo Point, NWT, 1986

Pennant, Thomas, *A Tour in Scotland, 1769*, Chester, 1771

Poole, Helen, *Lewes Past*, Phillimore, Chichester, 2000

Powell, Addison M., *Trailing and Camping in Alaska*, Hurst & Blackett, London, 1910

Rasmussen, Knud, *The People of the Polar North: a record... edited by G. Herring*, Kegan Paul & Co., London, 1908

Rider, Malcolm, *Hutton's Arse: Three Billion Years of Extraordinary Geology in Scotland's Northern Highlands*, Rider-French Consulting, Rogart, 2005

Robinson, Peter J., 'Ice and Snow in Paintings of Little Ice Age Winters', *Weather*, 2005

Schlombs, Adele, *Ando Hiroshige*, Taschen, Cologne, 2007

Scoresby, William, Jr, *An Account of the Arctic Regions: With a History and Description of the Northern Whale-fisher*, Archibald Constable & Co., Edinburgh, 1820

Seipel, Wilfried (ed.), *Pieter Bruegel the Elder at the Kunsthistorisches Museum in Vienna*, Milan, 1998

Stefansson, Vilhjalmur, *My Life with the Eskimo*, Macmillan, London, 1913

Thoreau, Henry David, *Walden (or Life in the Woods)*, Boston, MA, 1854

Tremper, Bruce, *Staying Alive in Avalanche Terrain*, The Mountaineers Books, Seattle, 2001

Weinstock, John, *Skis and Skiing from the Stone Age to the Birth of the Sport*, Edwin Mellen Press, Lampeter/Lewiston, NY, 2003

Acknowledgements

I would like to thank all the people I have met and interviewed in the course of travelling and writing this book for their assistance, which was often given freely and at short notice. For their advice and help I am particularly grateful to Laura Barber, Victoria Hobbs, Lisa Darnell, Aster Greenhill, Jonathan Radcliffe, Jane Birkett, Don Goodsell, Jonathan Jones, Ian Katz, and Ed and Jen Blincoe. Most of all, I would like to thank Lucy Blincoe.

Index